D0904472

PRAISE FOR
SOMEWHERE WE ARE HUMAN

"This is a book that makes visible those who have been invisible for years to the rest of the world. A wonderful and unprecedented collection [that] allows the reader to conclude, unequivocally, that at the end we're all human . . . the only difference is just a little piece of paper. Immigration is the new frontier in the struggle for human rights. If you really want to understand what it is to be an immigrant, to be forced to leave your home, and to arrive in a new country, you have to read this book."

—Jorge Ramos, journalist and news anchor with Univision

"This collection is not only a great read, but an important one. I applaud everyone involved."

—Luis Alberto Urrea, Pulitzer Prize finalist and bestselling author of *The Devil's Highway*

"This essential and moving anthology serves to humanize the dehumanized. The people who walk through these pages not only survive, but flourish; not only cross borders, but demolish the concept of borders; not only break down stereotypes, but build bridges, whether of language or steel. Even the title itself presents a challenge: if somewhere we are human, then everywhere we are human—even here, even now."

—**Martín Espada, National Book Award–winning poet of *Floaters***

"Urgent, necessary, and bold, the voices in this trailblazing anthology show us that their experiences are unique, but unquestionably part of the larger American story. Kudos to Grande and Guiñansaca for amplifying this vibrant community of artists and writers! Their meticulous selections offer us an extraordinary range of histories, perspectives, and—most touchingly—dreams."

—**Rigoberto González, American Book Award–winning author of *Butterfly Boy: Memories of a Chicano Mariposa***

"In this essential anthology, there lives both beauty and terror. So often these stories are told by others; now we get to hear them told by these artists themselves. What a gift as these vocal cords sing, ringing of human resilience and love, so much love."

—**Victoria Chang, award-winning author of *Obit* and *Dear Memory: Letters on Writing, Silence, and Grief***

"The voices telling these stories and the stories themselves have not always been embraced in publishing, which makes this book such a glorious gift to readers and so necessary and vital for our times."

—**Ligiah Villalobos, writer and executive producer of *Under the Same Moon (La misma luna)***

"Every piece in *Somewhere We Are Human* is so full of heart and rigor, it makes this collection one of the most important additions to undocumented literature. The poetry, prose, and visual art in its pages forge an incandescent testament of what it means to migrate, survive, and start anew."

—Ingrid Rojas Contreras, author of *The Man Who Could Move Clouds: A Memoir*

"A mesmerizing immersion into the lives of the undocumented, *Somewhere We Are Human* is a bold, definitive departure from the common and inauthentic 'study' of disempowered populations as portrayed by others. This brilliant and inspiring volume challenges us to create a world where it is not a sin to be born in places we must escape in order to survive, nor to travel to refuge in a place we are not welcomed."

—Carmen Tafolla, State Poet Laureate of Texas and author of *I'll Always Come Back to You, Arte del Pueblo*, and others

"The only anthology edited by and featuring primarily writers who've been undocumented, this glorious collection speaks against the power of the state and what the state can't see: the fullness of people who are so much more than their papers. A convening of voices from across the global south, these writers testify of the pain of family separation and the classed experience of migrant labor, also of organizing against ICE, prisons, and the DAPL pipeline, and also of queer love, the joy of clubbing, and the love we have for those who came before us. This is what solidarity looks like!"

—Ken Chen, award-winning poet of *Juvenilia*

"*Somewhere We Are Human* incites in me the kind of riot of heart and feeling and thinking that can only happen when many voices are sounding—all at once—part of the history of the world. Vital and radiant. It is a book so plural and shiftful that you cannot be the same after reading it."

—Aracelis Girmay, award-winning poet of *Teeth, The Black Maria, and Kingdom Animalia*

SOME WHERE WE ARE HUMAN

AUTHENTIC VOICES ON MIGRATION, SURVIVAL, AND NEW BEGINNINGS

SOMEWHERE WE ARE HUMAN

EDITED BY

REYNA GRANDE

AND

SONIA GUIÑANSACA

HARPERVIA

An Imprint of HarperCollins*Publishers*

Some names have been changed to protect people's privacy.

Excerpt on page 175 from Tuck, E. T., & Ree, C. (2013). *Handbook of Autoethnography.* S. L. Jones, T. E. Adams, & C. Ellis (Eds.). Anchor Books. Excerpt from "A Glossary of Haunting" reprinted by permission of Anchor Books, an imprint of Knopf Doubleday Publishing Group a division of Penguin Random House LLC. All rights reserved.

HarperCollins books may be purchased for educational, business, or sales promotional use. For information, please email the Special Markets Department at SPsales@harpercollins.com.

FIRST HARPERVIA EDITION PUBLISHED IN 2022

Designed by SBI Book Arts, LLC

Dandelion and seeds illustrations by onecentdesign/Shutterstock
Page 9: Where do we go *(2018) by Bo Thai, printed with permission*
Page 28: Courtesy of Jennif(f)er Tamayo
Page 38: A Future, Elsewhere *(2020) by Alan Pelaez Lopez, printed with permission*
Page 169: A Moment for Two *(2021) by Julio Salgado, printed with permission*
Page 252: Come Back *(2020) by Rommy Torrico, printed with permission*
Page 276: Luchadora con fe todo se puede *(2017) by Miriam Alarcón Avila, printed with permission*

Library of Congress Cataloging-in-Publication Data has been applied for.

ISBN 978-0-06-309577-9

22 23 24 25 26 LSC 10 9 8 7 6 5 4 3 2 1

Yaccaira de la Torre Salvatierra,
this one is for you, mujer!

—**REYNA**

To my parents, and my abuelitos—
este libro y todo es siempre para ustedes.

—**SONIA**

CONTENTS

Contents

Contents

Contents

Contents

FOREWORD

Human beings have always migrated and will certainly move even more in the era of climate catastrophe, but the United States of America has laid claim to the idea of migration as part of its ideology. This way of thinking proclaims we are a country of immigrants. But don't call it ideology. Americans believe that ideology is for the Marxists, socialists, communists, and critical race theorists. The American Dream exists beyond ideology, as the dream everyone all over the world must be dreaming—according to Americans.

Immigration into the United States by certain people terrifies some Americans. It could also be argued that the idea of immigration validates America to these same Americans. Of course, people want to come here, for we are great, or will be great again as soon as we have the right kind of people coming here. This anthology, *Somewhere We Are Human*, takes on the paradox of immigration and xenophobia that exists at the heart of America, by which I don't mean the United States in reality, but this mythical country of the American Dream that is so deeply rooted in the American psyche that many Americans, even the liberal and critical ones, have a hard time unrooting it.

The stories, poems, and artwork in this anthology address this paradox wherein America represents itself as the land of "new beginnings" but is also a place that is so demanding of its newcomers that they must think, "Somewhere we are human." If America is great, then why is the humanity of anyone in question? Why must anyone long for their own innate humanity, one that is denied by some and even by many? The American story of so many newcomers—as well as indigenous people and the descendants of the enslaved—is therefore one of "survival."

Racism, indifference, misunderstanding, microaggressions, exploitation, family separation, the terror of being undocumented and under threat of deportation—these are all part of the terrain of danger and survival for newcomers. In this sense, the anthology affirms what is already well-known about the American story of migration, which is that in order to become American, to participate in the American Dream, to be a part of American exceptionalism, the newcomers often and unfortunately have to undergo rites of initiation that range from contempt to brutality.

Part of the American mythology is that these regrettable aspects of the American experience will ultimately be overcome, by individuals and the nation as a whole. This anthology challenges that optimism, as Jesús I. Valles expresses succinctly in his poem "you find home / then you run," when he writes "i do not have a country." Composed in the era of "Make America Great Again" and its aftermath, or perhaps simply its fallow period before it rises once more, the anthology's mix of pessimism, defiance, and occasional optimism captures the mood of so many Americans. Many, on all sides, are worried about the fate of the country. For some, this is wrapped up in the idea that what America represents—the city on the hill, a beacon for the world—is under threat. For others, including many in this anthology, the worry is about how to live in a country that is simply, for better and for worse, home.

One of the most unique things about *Somewhere We Are Human* is the visible presence of those writers who are both geographically mobile and who have moved across norms and boundaries of sex and gender as queer, nonbinary, or trans. This is not accidental. The borders of nations are not only racialized, politicized, and militarized, but also gendered and sexualized, as Gloria Anzaldúa pointed out in *Borderlands/La Frontera.* And the sociologist Aníbal Quijano has argued that America, as the continental region from north to south where contemporary capitalism reached its apotheosis through colonialism, required the construction of a bourgeois, heteronormative family. This is why the proponents of "Make America Great Again" cast unwelcome migrants as murderers and rapists since both threaten this kind of family, which also stands in for the nation. And this is why an anthology that is ambivalent about America itself throws the borders of geography, nationality, gender, sexuality, and identity into question.

While we have no shortage of immigrant stories, we have not heard as much as we need to from the undocumented and previously undocumented who comprise this anthology. Their stories drove home for me the perpetual feeling of dread that hangs over so many of the undocumented, as well as the power and necessity found in their courageous voices. While these writers may be undocumented in a legal sense, they are documenting themselves, and this country, through their writing. May this anthology be a cornerstone of an undocumented literature that galvanizes our collective conscience and imagination over what is possible for this country.

Ultimately, *Somewhere We Are Human* asks an implicit question: Where will we be human? Is it in the United States? Is it in this dream of a mythological America? And if our humanity cannot be realized in these places, then why not, and how can we make that necessity happen? The answers can emerge only through struggles over justice and equality that will not be resolved anytime soon. In the meantime, the defiant claim of féi hernandez's poem "After Sappho" resonates: "America was always mine."

Viet Thanh Nguyen,
Pulitzer Prize–winning
author of *The Sympathizer*

EDITORS' NOTE

Dear Reader,

We are honored to present to you these powerful stories of resilience, tenacity, and hope. Our younger selves never imagined putting a book like this together—a unique collection of essays, poems, and artwork by and about people like us: undocumented and formerly undocumented immigrants. Since we are part of a community that is often relegated to the margins of society, where so many of us are forced to live in the shadows and where our voices go unheard, the publication of *Somewhere We Are Human* is to us a dream fulfilled. This book is a celebration of the unbreakable spirit and incredible talent of our communities.

We both migrated when we were very young, Reyna from México and Sonia from Ecuador, and landed on opposite coasts, California and New York, respectively. We both grew up undocumented and faced similar challenges as we struggled to find our place in the United States. Luckily for us—two brown girls living in Harlem and Los Angeles—we found a sort of sanctuary in our public libraries where we discovered the power of stories and developed a fierce love of reading. Unfortunately, the books available to us too often failed to include experiences like ours. We craved narratives that spoke to our reality as immigrants and the complexity of being undocumented; stories that reflected our humanity, where we felt seen and heard.

With this collection, we hope to contribute to the ongoing and evolving conversation about immigration policy and justice by centering authentic stories of immigrants. Never has the need for such a book been greater, and we gave ourselves over to the writing and editing of it as if our lives depended on it. Maybe they did. After all, the creation of *Somewhere We Are Human* happened during a time when DACA—the Deferred Action for Childhood Arrivals in the United States—was up for debate; during the ongoing deportation of migrants across the country and the imprisonment of people in detention centers; during an increase in anti-LGBTQ policies; during a rise of xenophobia and anti-Asian violence; during the uprising for Black lives; during an election; during a climate crisis, economic instability, and a global pandemic.

At a time of so much unrest, how do we shift the nation's collective imagination about migrants toward one rooted in humanity and justice? There were many questions that guided us as we selected the essays, poems, and artwork you will find within these pages.

The process to arrive at this collection was done in two phases. First, we created a wish list of people we wanted to include—prominent voices doing groundbreaking work in immigrant activism and cultural and literary spaces. We also wanted to spotlight new immigrant voices from across the country and those who no longer reside in the United States because they decided to leave or were deported. So in addition to seeking specific contributions, in October 2020 we launched an open call for submissions online, which closed in December of the same year. The response was overwhelming but not surprising. Opportunities and infrastructure for immigrants to tell their own stories in their own words are few and far between. What we ended up with was a pool of submissions about humans who love; who have complicated relationships with family members; who are engaged in multiple social justice issues, such as climate change and reproductive and LGBTQ rights; who are figuring out their connection to their homelands and their adoptive country; who are exploring their identities, pursuing their dreams, and owning their truths.

In this collection, we wanted to go beyond the data points—the case studies or policy reports—that all too often dominate mainstream discourse. The pieces of the people you will meet in this collection are never-before-published, firsthand accounts of lived experiences. These are authentic voices of emerging or established writers and artists, of leaders in their professions or communities. Some of us have been granted Temporary Protected Status or live under the DACA program. Some of us are in between status adjustments; others are deportees, in exile, or are waiting for asylum approval. Some of us have green cards or are now American citizens, and some of us do not qualify for any current relief efforts because we have aged out or are awaiting a court hearing, stuck in the ever-increasing judicial backlog. This range was important to us because it puts on display the settler colonial policies of immigration and the violent system that we are all a part of, against our will and sometimes complicitly.

These government-made statuses cannot fully capture the reality of human lives, and the contributors in this book are trying, individually and collectively, to make sense of the gray areas by revisiting our memories and learning how class, race, religion, sexuality, gender, nationality, statehood, ability, and language have shaped our lived experiences in this country. We broach subjects often not included in larger mainstream migrant discourse, like mental health and body dysphoria, and topics like shame and guilt that we often don't speak out loud. We explore our internal conflict of trying to hold on to our roots while at the same time trying to fit in; of wanting to have a better life while knowing the price we must pay is too high; of living a reality that, more often than not, fails to meet our expectations; of having to fight, again and again, for our right to remain and make a home here. We make art, we create, we write to make sense of it all. To heal and make ourselves whole again.

Somewhere We Are Human is not complete. One book can never encompass the multitude of migrant experiences. But this book is unique in that it was created exclusively by undocumented or formerly undocumented immigrants—from the compilation, editing, writing, and art to the beautifully embroidered dandelion on the cover. Although the monarch butterfly has become a symbol of migration as an act of survival, we chose the dandelion to symbolize our journeys. Everyone loves butterflies. The same cannot be said of dandelions, which, regardless of their beneficial properties, are considered weeds. Too often, immigrants are also seen in a negative light. Luckily, dandelion seeds are strong and adaptable. We've all seen them ground and bloom in the most unlikely places. Like the dandelion seed, with the odds stacked against us, we immigrants do our best to make new lives and continue existing wherever we land. This is why we organized the collection into three sections defined by the phases of the immigrant journey: migration, survival, and new beginnings—a journey not unlike that of the dandelion.

The title *Somewhere We Are Human* was inspired by a line from one of Sonia's poems. It reflects the humanity of each individual who has been forced to abandon their homeland and seek a new life. Living in a country that often denies our humanity takes a toll, but this anthology is a declaration that even if we are not seen as human beings in the city, the county, the state, wherever we live, we *are* human everywhere, and no one can take our humanity away.

We see this collection as an offering to our migrant communities but also to those on the outside wishing to know more. If you are an immigrant, we hope you see this book as a celebration of who you are, that the stories within these pages make you feel seen, heard, and less alone. If you have ever left your home, either willingly or by force, we hope you join us in finding home together as we pursue a future not defined by borders.

Every migrant has a story, a life, a family, a face, a voice. Like the dandelion, our stories are the seeds that carry life and offer new beginnings wherever they take root.

In solidarity,
Reyna and Sonia

MIGRATION

Sonia Guiñansaca

POEMS

Before

Starts with a morning ritual
Mama Michi braids my mom's hair
Pouring cold water from a chipped cup
A baptism over the bathroom sink
The comb is missing teeth
There is a parting
And weaving
Of hair strands
Into blessings
Only abuelas
Know how to make
My aunt, Rocío
Waits her turn

They hurry off to school
Clumsy
Bellies full of gelatina

Mada! Chio!

A classmate calls out from the playground
They run towards each other like children do
Mochilas clapping on their lower backs
Rose cheeks tinted by an equator's sun
Erupt into giggles

And when they become teenagers
The 80s crosses over
The southern border
In lycra

Rocío's hair becomes as high as the sun
Standing firm with mousse
Using the bowl of the spoon
Mami learns to curl her eyelashes
Bending with metal
A sort of magic
Only they can find in el campo

Somewhere
They were this glorious before

Somewhere
They've always existed
Before the migration

Sonia Guiñansaca

Reunion

Dad left first

Mami leaves when I turn one

Takes them four years to save up money

I come by airplane *Safely*

I'm five and arrive with resentment

When I see Dad, I call him Rodrigo first

When I argue with Mom, I remind her *she wasn't there*

At sixteen
I learn to pinch Google Maps to find the border. It tells me that it is 3,048.9 miles from Ecuador to NYC *by flight*. Does not calculate the steps of two adults who walk from Cuenca to Panamá to Guatemala to México to Texas while holding liters of water. Does not measure the length of rivers Dad crosses in old tires while carrying a Manhattan address written on an envelope. Tucked deep into his pocket. Creasing with every step. There is no way to quantify the amount of crawling Mami had to do on dry soil between desert bush and panic. No map outlining where they called out for God during drownings. Does not point out where la migra's helicopter is hovering like a scorpion with wings

They both survive   part human and part miracle

When they finally tell me their journey
I keep saying *why did you leave me*

when I mean *why didn't I come with you*
when I mean *I've missed you*
when I mean *I'm sorry*

Sonia Guiñansaca

After

Like gold a *good immigrant* doesn't tarnish
Like gold we are extracted and polished

I shine on a magazine cover
Mami cleans the same colleges I perform at

Papa Jerry is told to extract
The last gold tooth he got in Ecuador

Wearing his new dentures
Papa Jerry can't return to bury his parents

He grinds his teeth at night for fifty-one years
And keeps digging

I'm told to wear this green card across my neck
Like a gold chain spelling out my name

And then
After we become gold *what do we dig for*

As children we had dirt under our nails from countries
we undug

After the
socialsecuritynumbersthepapersthestatusthejobthedream

don't our hands hurt

Maybe we don't want to be like gold
Maybe buried deeper   *somewhere near our elders' feet*

Maybe we are tired
Maybe I want to be   earth   *human*   *ash*

Bo Thai

Where do we go (2018)

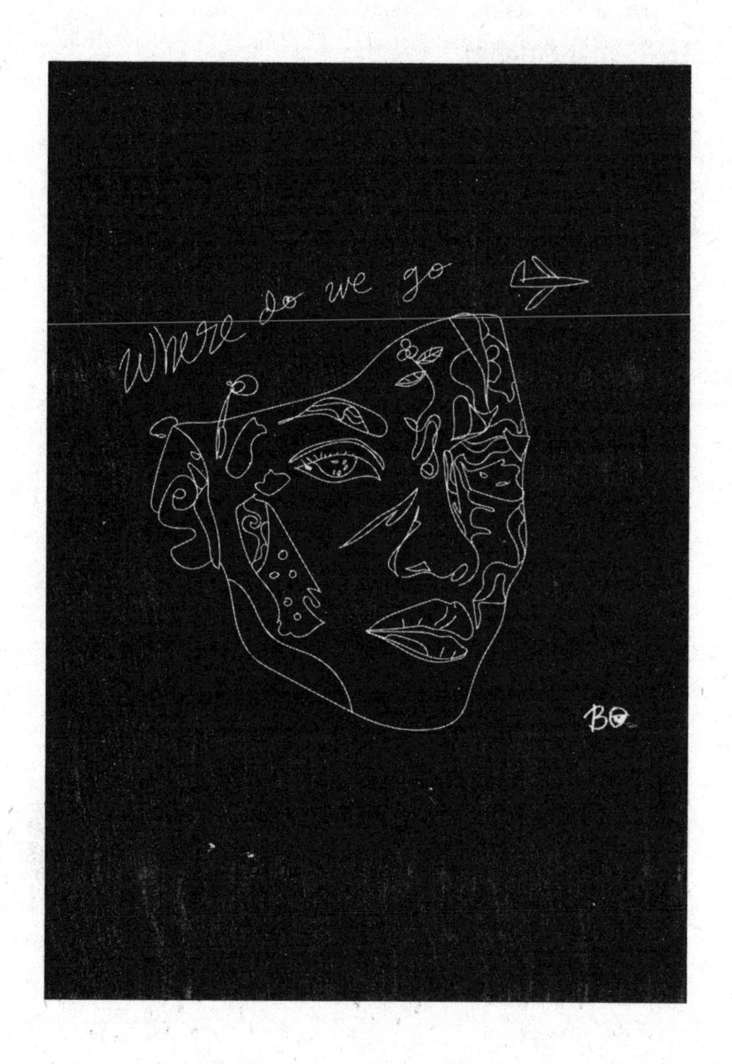

Bo Thai, or Boonyarit Daraphant, is an undocumented artist. His work ranges from writing to visual arts to clothing design. Without DACA and therapy, he initially got into art as a form of healing and as a part of his hustle to create two clothing lines: *illegal Drip* and *BoThai*. His hope is to use art to hold vulnerable spaces for himself and others. He often freestyles his streams of thought and later refines them for larger audiences. His creative inspiration comes from his culture and experience in Thailand, anime, surrealism, and other mediums.

Carolina Rivera Escamilla

The Promise Made

I had not expected Mamá and Papá to support my leaving El Salvador. I was afraid they would be angry or disappointed with me for having gone behind their backs to seek political asylum in any country that would be willing to accept me. I brought out the paper I had hidden from them for a whole day, convinced they would never allow me to go. I handed it to my mother nervously. "I received this from the Costa Rican UNHCR office."

"What is this?" Mamá asked.

"A telegram that says I am accepted for asylum."

"How did you get this telegram? How do they know you? Why wasn't it sent here to our address?"

"It was safer to use another address, Mamá."

One by one my brothers, sisters, and I were becoming exiliados, immigrants, refugees.

More than a year and a half had passed since my high school graduation. Political and economic forces were strangling me, my family, the whole country. I had to leave El Salvador. I wanted to stay, but I could not. I had barely managed to graduate from Centro Nacional de Artes, my theater arts high school, especially after soldiers had looted, destroyed, and plundered it at the end of my first year. That same year in November 1980, civil war broke out.

My life as a high schooler soon became about missing a class in favor of a protest, attending meetings to learn how to create banners and use aerosol spray cans to graffiti important walls in the city to denounce the military's repressive violence against the people, especially against us students and our families. We lived through Archbishop Oscar Arnulfo Romero's assassination and a massacre at the University of El Salvador, where students were bombed and burned alive. The first school year ended with rape, murder, and the discovery of the buried American nuns almost within sight of the national airport. Then the first guerilla uprising led to the military leaving bodies of young people in the city street gutters for all to see. When my high school was raided and wrecked, I got more consciously involved in the struggle because students and teachers were taken away and disappeared, but no one could say a public word about it without fear of repercussions. Finally, after the school was shut down as a result of the damage, the ministry of education scrambled to find us safe locations where we could finish our education. We students became nomads, our classes relocating from one place to another: in the mini-theater within the National Theater, in an empty storage room within the National Symphony Hall, in a big concert hall within the Department of Music. Once, we even held class in a park. Nowhere was safe for us. We were shifted around in the heated chaos swallowing up the country.

Then, early one morning, my sister and I were held up at

gunpoint at the National Stadium while she was training for a race. Men with dark glasses tried to push us into a van, but we managed to escape from them. They might have been the same death squad that had pursued my older brother to our home months before. As they pushed their way through the front door, he had escaped through the back two minutes before.

Later that month, my sister and I met a doctor on the stadium running track. She gave me the address of the Costa Rican Consulate and encouraged me to request political asylum. "You both should go," she insisted, but my sister said, as the oldest daughter, she needed to stay with our parents. I told no one I would write the letter, not even my sister. I knew my parents would never let us go. Mamá had once said, *"I would rather that you die here among all of us, rather than be raped, killed, or disappeared somewhere else. At least here, remember, we found your cousins' bodies to bury."*

I had taken a gamble and it paid off. Mamá put the telegram down and surprised me when she said, "It's better, daughter, that you go. Better for you to be safe there than to get tortured or killed, tossed aside like a dismembered doll by a death squad. Look what happened to your cousins. I'm relieved your brothers have already gone from here."

That night, lying in my bed staring at the corrugated Duralita ceiling thinking about my situation, I realized I was about to leave everything I knew: friends, family, the disappeared, the fleeing, the dead. Most of all, I dreaded losing my family, my home. I would miss everything associated with them, like yuca frita, pastelitos con carne, sweet quesadillas horneadas, or the occasional weekend pupusas. Would I find these tastes of home, always shared among family, in my new haven?

I tried to imagine a new life in that safe place. I knew for sure I wouldn't find Doña Amalia's pupusas. The smell of sunzas, mangoes, and stinky pigs bathing in the rainstorm puddles on broken paved streets would not be over there. November's edible pink summer flowers on the maquilishuat trees would not shed pink petals on my head. The street dogs would not chase me when I ran uphill to catch the bus. The color of the sky, the perfume of familiar places . . . the National Theater, the University of El Salvador, the National Library, the echo of friends' laughter, the embrace of family would not be over there.

The afternoon before I left, I walked to the magazine stand and picked up an old magazine with the Rocky Mountains on the cover. The headline read: "Journey to USA and Canada Today." Mount Alberta was all dressed in snow, like a massive shaved ice cone piled high. I promised the magazine salesman I would pay him eventually or bring him another old magazine. He did not know, and I could not tell him, that I was leaving. He told me he liked *Life* magazines, especially with blond women, like Marilyn Monroe, or music groups, like the Beatles, so he agreed to the deal. I am still in debt to him. As I walked home, I looked at the headline and made myself believe I was going on a grand voyage.

"See," I said to my parents, flipping through the pictures in the magazine, "these are nice places!" I was trying to make us feel better. These cold places were no more real to me than the skinny, tall model covered toe to head in ski clothes and equipment, like an alien in a space suit. We had found out earlier that day that my asylum would be somewhere in Canada. The United States was taking no war refugees from a country whose military they supported. Mamá cried and, while holding my hands, made me promise that I would join my brothers in the US as soon as I could. I promised Mamá that, one way or another, I would find my way to them.

On June 18, 1985, I stepped onto a plane. I did not know whether I could ever return to my family, to my country.

During the four years I was in Canada, the promise I made Mamá hung on my ears like weighty earrings. I applied for a tourist visa to go to the US to visit my brothers. When I finally arrived at the airport in Los Angeles in 1989, passing through the immigration agents, the image of robotic cops came to mind from movies I had watched before landing in Hollywoodland. The massive concrete construction of the airport intimidated me. But my fears faded as soon as I saw my brothers and my cousin, each extending bouquets of yellow, red, and white roses to me. I hadn't seen them in more than five years. We cried, we laughed, we hugged right there outside baggage claim. At their apartment, we talked for a long time, remembering people we knew in the colonia, wondering whether they were all right. We went to sleep at three in the morning.

In Los Angeles, I was told again and again how lucky I was. "Lucky you took an airplane to Canada and then to here!"

Most of the people I was around were undocumented. I heard the stories of my brothers, my cousins, my compatriots, of their escapes and multiple border crossings. They'd entered the US hidden inside impossibly small false compartments in trucks or inside trunks or under backseats of cars. They crawled through sewers. They spent nights in mountains or near rivers waiting for the right moment to cross the border. The stories of injustice compounded on their journeys north.

By the end of the first week, without anyone's permission, I went out alone to explore the neighborhood. I ended up far down at the intersection of Western and Wilshire Boulevard in this place of seemingly endless avenues, streets, boulevards. Incredibly tall palm trees poked the great blue sky over the city of Los Angeles. There were too many business signs that gave me vertigo, too many cars speeding by. I rarely crossed paths with another pedestrian. Isolation and loneliness made me wonder if staying in Los Angeles was the right thing. Was never returning to Canada the correct decision? Mid-thought, the image of Mamá holding my hands came to me, with me promising again, *"Yes, Mamá, I'll find and stay with my brothers."* An echo of her approval bounced off traffic noise, affirming this decision at the intersection of Western and Wilshire. That was the moment I resolved not to return to Canada, to start a new life near my family. I let my US tourist visa expire, and just like many undocumented immigrants before me, I bought a fake green card at MacArthur Park.

A small nearby Salvadoran restaurant became a favorite place to eat, sometimes with my brothers, sometimes alone. The owners from El Salvador's countryside had just recently opened it. The place reminded me more of an indoor comedor in San Salvador, a disorganized, impromptu kitchen where working people grab something inexpensive to eat. I liked it. I enjoyed eating real beans, plátanos fritos with queso duro-blando and crema. I would eventually eat every type of pupusa they had to offer. Four years of eating unsalted Canadian bacon and eggs led me to this unkempt heaven. Everyone there talked about El Salvador. With only three tables in the tiny place, we all huddled together into one huge conversation.

Filled with uncertainty and rage, we wondered how this war would end, how we might find normalcy again. As civil war in our tormented El Salvador continued, and thousands were killed and tortured, we contemplated our mixed blessings.

Carolina Rivera Escamilla is an educator, writer, poet, and filmmaker who lives in Los Angeles, California. Born in El Salvador and educated in theater arts, she went into exile in the 1980s. Her writing has been published, among other places, in *Analecta Literary and Arts Journal* (University of Texas, Austin), *Hostos Review/Revista Hostosiana* (Latin American Writers Institute, CUNY), the *Strange Cargo: An Emerging Voices Anthology 1997–2010* (PEN Center), and *Collateral Damage: Women Write about War* (University of Virginia Press). Her book of short stories, entitled . . . *after* . . . , was published in 2015. A fellow of the PEN America/Emerging Voices program, Rivera Escamilla was also the director, writer, and producer of the documentary film *Manlio Argueta, Poets and Volcanoes*. She earned her bachelor of arts degree in English literature at the University of California, Los Angeles, with an emphasis in creative writing and Spanish literature.

Jennif(f)er Tamayo

& I Came the Way the Birds Came.

I

McAllen, Texas, is well-known for birds. The nine locations of the World Birding Center—a "World Class Destination for Birders!"—run along the edge of the Rio Grande, a flowing body of water where all kinds of migratory birds stop to rest during their north-south crossings. The World Birding Center, its website boasts, is "where you will *truly* find 'A whole New Nature Adventure.'" The tab on "Bird Information" has a grid of birds and their respective calls. I look the most like the ████████ but I sound most like the █████. And Ringed Kingfishers—with their shaggy blueblack crests—have always been my favorite because they look femme and regal, and their eyes make me nervous.

II

In spanish the word *pájara*
the word for female bird
is also used as a derogatory slang
for a lesbian.
As in, *¿ella?*
. . . *ella es bien pájara.*

I learned this from another website.

III

My mother and I made our crossing at the very location where the birds make theirs, through the Rio Grande, no more than a few miles from the World Birding Center's Hidalgo location. From what I've been able to piece together, ██████████ we made it to ████████ somehow and then, after █████████ under a car, crossed into ██████ with the help of another animal, █████████. The path we took from Colombia through México and into the US follows the Mississippi River flyway—a migratory route famous for its lush deltas, rivers, and marshes. This region's waters—the *Angry* River, the Mississippi, the Gulf—hold knotted histories of flight and escape. In the mid-1800s, Black captives used the Rio, the "bird paradise," as a site of fugitivity. Which really means that: even before I was born, even before you were born, Black people were making pathways for our safety.

IV

The last time my mom let herself
tell me about our crossing
was a few years ago;
mid-sentence,
while on another topic,
she randomly rememb-

the *flash*
a lightning storm, scratching out our sky

¿Te acuerdas?

In these moments, she is ready
to let some of the story
leave her body
to take flight.
Her eyes kinda roll
upward toward the sky.
When we talk about it
—and we rarely do—
I turn into a chick
in her lap, taking in memories
bits of wet, ragged worm.

V

The World Birding Center is a place for families: bird families & white tourist families. Admission is $5.00 for adults, $3.00 for seniors. Kids are free. I haven't been back to that area since my mother and I were detained at McAllen's other famous structure: the Border Patrol Station just a few miles down the road. Built in 1921, this particular Station is one of the largest in the region, patrolling fifty-three miles of the Rio Grande border from "terrorists and weapons of mass effect." The Station's jurisdiction intersects with many of the World Birding Center's numerous "bird" watching towers. It was at this Station where a border patrol officer told my mother to leave me, her four-year-old child. *Better for both of you to be separated. Better treatment*, he told her, and then, like a villain in a cheesy movie, winked.

When she releases this part of the story, when it hovers in the air between us, I feel a lightness inside me, as if I too would   lift   from
the ground   and   fly   away

VI

This past year, my mother gifted me the wool sweater she wore when we crossed: an '80s big puffed-shoulder cyan and hot pink feathery soft thing given to her by my father who stayed behind in Medellín. Correction: gifted to her by my father who was *able*, who felt safe enough in his body to stay in Medellín. He was ████████████████████ ████████████ but ██████. She tells me the story little by little, years, decades at a time, generations at a time—giving me the sweater is part of it. I think about what the birds wear when they migrate. What they carry with them besides a full belly. My mother brought the soft cyan sweater and a baby spoon for curling her eyelashes. And her child. When I wear it, its wool fibers tell me all kinds of stories. When I wear it, I feel *bien pájara*, ready to puff my crest, to be unruly and wild, ready to revenge.

VII

According to this one review on TripAdvisor, the best time to visit the World Birding Center is April. "Don't go in January," says one user. "NOT worth it this time of year . . . We see more birds in our backyard at home . . . VERY disappointed with this attraction." I think of this user and her family (from ████████, by the way!) with their stupid little binoculars trying to catch a glimpse of a kiskadee from the edge of a watchtower. This *place* is their "attraction," their amusement park. I imagine them in their North Face puffer vests or camo gear. Looking. Looking. Looking at the lush valley . . . when suddenly the Prothonotary Warbler, a bright little songbird, flies beak first into their rotten eyeballs. The user shrieks, the kiddos caw, and a little trail of blood follows them to their Subaru.

VIII

More Google searches reveal, unsurprisingly, that the U.S. Fish and Wildlife Service, part of the U.S. Department of the Interior, is one of the World Birding Center's key business partners. In fact, it was the Department of the Interior that funded their green-eco-friendly "award-winning" and—in my opinion—ugly ass tin can buildings. "Over 10,000 acres will be opened up," says the "About Us" section. "Bird watching is a booming national pastime . . . it's also B-I-G business." The World Birding Center is basically America. Acts of violence hidden neatly inside the language of care and prosperity. Choreographies of extraction, of land theft, of surveillance and policing disguised as a hobby, preservation, and ecotourism. For a small fee, *you too* can come patrol the border!

> Sometimes
> most of the time
> these motherfuckers are so transparent,
> I can't stand it.

IX

While I am writing this
a call from Medellín
enters:
my father.
I don't pick it up

because it's ██████ to ███
& I speak bird now.
I can only
squawk
and shrrreet shrrreeeet
and caw-caw
and speak bird.

X

Hovering over the shoreline, kingfishers have a particular eye, their own unique browser histories.

The first and only time I've seen one, I heard her screaming call before I caught her dark oily eye in the branch. Her call cut through the space, and I could feel she was really strong, like maybe *too* strong. I wonder what bird calls we heard when we crossed; did kingfishers call out to me and my mother as we moved through the river? What were they trying to say to us? What did they warn?

Or, maybe, we were the warning.

XI

It is not only wool sweaters and baby spoons that non-Black, non-Indigenous migrants like me and my mother bring with us. We bring our selective colonial memories, our bootstrap habits, our antiBlackness; whether we like it or not, our crossing bodies participate in the process of settlement on both sides of this river. During COVID-19 quarantine, Afro-Femicides have been on the rise in much of Colombia. Women are literally caged inside the violence of patriarchy, of whiteness, of the State. Black Colombian women, Indigenous women are subject to a particular dispossession that migrants like my mother and I too easily lose sight of when caught in the trauma of our own crossing. I mean, would the officer have suggested separation to a Black mother or just forced it? Would the wink have been something more gruesome to a mother who spoke only Nahuatl?

I want to be the kind of bird that can love better, can love with more ■■■■■ inside of all the violence.

XII

When I'm so angry, *so* angry—

the flash
the lightning storm, scratching out our sky

No. No
me acuerdo

XIII

I force myself into a prolonged Google/Wikipedia spiral into the year 1921: the year of the Tulsa Massacre in Oklahoma; the year the United States and Colombia decided on the fate of Panamanian peoples by drawing colonial lines across our maps; the year congress passed the Emergency Quota Act to restrict "undesirable" immigration to the US; and the same year the McAllen Station was built from the ground up by just four officers. *100 years, the length of a long healthy lifetime.* And then I think that long before 1921 the Rio Grande, by other names, flowed through the Tāp Pīlam Coahuiltecan Nation. And, for many lifetimes before and many lifetimes after, that river will continue to flow. And the kingfishers, and cowbirds, and orioles will make their crossings at these spots. The bike trail will be disappeared by blades of green. The "Hawk Tower" will decay, grow covered in that pearl-colored bird poo. And the water protectors will dismantle the dams and watersheds that have interrupted the river's body. And that sovereign río will flow and those birds will loop around the skies as they've always done. And this shit, this American blight will all be over. Or, at least, become something else.

XIV

And maybe I'll become something else too. A new kind of bird, an unearthly bird. My friend Laila, who lives between El Paso and Ciudad Juarez, sends me a text message: "You are a beautiful bird." Can you see how the sleeves of my shirt kinda look like the beginning of a wing? Maybe I'll transform into a cygnet, feral and soft and queer. *Bien fuckin pájara*. I want to be bad, like the Bronze-Headed Cowbird, who will do anything for love. And did you know that the Groove-Billed Ani lays eggs in a communal nest? Imagine the warmth of that brood; the untrained dreams those little chicks must have together. Sometimes, I want to be a monstrous

bird with a bill so whetted it rips across space-
time:

to a place where flesh is not a border between
life and death. Like the blessed Rock Pigeon
I want a long fucking memory
an ancestral grudge kinda memory.
I want us to remember that
it has been Black and Indigenous kin
who have taught us—over many lifetimes—
how to take flight. How to orient ourselves
toward freedom.
And the land,
it was her
who made our journeys possible.
We must transform
into something
else, something more ████████
and more ████████
if we are to plot
our next ███████

Author's Note: Details and direct quotations about the World Birding Center can be found at www.theworldbirdingcenter.com. The Bentsen–Rio Grande Valley World Birding Center headquarters received a 2006 American Institute of Architects Top Ten Project award—and cost "the state" of Texas $3,500,000, not including the land itself, which continues to be owned by the settler-colonial empire. More on the history of Black captive escape through the southern border can be found in Russell Contreras's article "Story of the Underground Railroad to México Gains Attention," published by Associated Press News on September 16, 2020. The brief history of the McAllen Station and the phrase "terrorists' weapons of mass effect" can be found on the Station's U.S. Customs and Border Protection website: www.cbp.gov.

Jennif(f)er Tamayo is a performer, poet, and scholar whose writing and stage work reimagine narratives about and politics of queer, undocumented figures in the U.S. She is the author of the poetry collections *[Red Missed Aches, Read Mistakes, Red Mistakes, Read Missed Aches]* (2011) and *YOU DA ONE* (2017), and her latest prose-poem publication *to kill the future in the present* (2018). JT has received fellowships from the Cynthia Woods Mitchell Center for the Arts, the Arts Research Center, the Hemispheric Institute, and CantoMundo. Her work has been staged at the Brooklyn Museum, BAMPFA (Berkeley Art Museum & Pacific Film Archive), Midtown Arts & Theater Center Houston, and La MaMa Experimental Theatre Club, among others. Currently, JT lives and works on Ohlone and Patwin lands and is a PhD candidate in performance studies at the University of California, Berkeley. Her research explores how contemporary Black and Indigenous poets use vocal and sonic practices to counter-narrate histories of colonial violence.

Javier Zamora

POEMS

Every election,

a candidate promises: papers,
papers, & more.
They gift us Advance Parole.
We want flight. "OK, then,
make some wings."
We do. Some of us on the edge . . .
¡leap! Some watch others fly
& are scared. Others
are too old for anything . . .
The candidate wins,
the candidate loses, we fill more forms.
Consider marrying our friend's cousin
whose business is 20k, 30k,
the easy way. We don't—
because we think
flight will come one day. That day
someone will finally see
we're not made of blue,
or black signatures alone.

There's a Wall ft. Merengue Legend Kinito Mendez's "Cachamba"

There's a wall (there's a wall),
there's a wall where people are tanning, Speedoes, bikinis,
inside of a hammock
on top of the wall near the Mexican Desert.

There's an Agent (there's an Agent),
there's an Agent (Xicanx) running
toward the people with Speedoes, bikinis,
inside of a hammock on top of the wall at the edge of the
Mexican Desert.

There's a Deputy (there's a Deputy),
there's a Deputy who ordered the Agent
"¡Arrest the people! in Speedoes, bikinis,
inside of a hammock on top of the wall at the edge of a river
in the Mexican Desert."

A Commissioner (a Commissioner),
Commissioner who said "¡No!"—
Secretary who said "¡Yes!"—
to the Deputy who ordered the Agent running
at the people tanning, Speedoes, bikinis, inside of a
hammock
on top of the wall at the edge of a river in the Mexican
Desert.

There's a Secretary (The Secretary),
Secretary who told Commissioner "I'm in charge."
Commissioner who told Deputy "¡No!"
who told Agent "¡Arrest!"
the people tanning in Speedoes, bikinis, inside of a
hammock on top of the wall
at the edge of a river in the Mexican Desert.

The President (Ohhh *The* President),
President who said
"You're fired" to
The Secretary who said "I'm in charge," Commissioner
who told Deputy "¡No!"
who sent Agent running
toward the tanners in Speedoes, bikinis, inside of a
hammock on top of the wall
at the edge of a river in the Mexican Desert.

¡& the wall is me! (¡Yes, it's me!)
I'm wearing a Speedo (¡Yes, he is!)
Inside a bikini (¡Yes, he is!)
I'm not Mexican (¡No, he's not!)
¡Let me tan please! (¡Let him be!)
Please let me breathe (¡Let him live!)
Fuck the B.P. (¡Yes, fuck them please!)
Así, así, así, mamacita, así.
Así, así, así, mamacita, así . . . y ya.

At the Naco, Sonora Port of Entry Twenty Years After Crossing the Border, but This Time with Papers

"Research,"
I tell the Mexican border guard
who stamps my passport.
It's my first time back
in Sonora. I want to find
the exact route I took from Hermosillo
to Naco, the coyote hideout,
the albergue, to feel closer
to those who were with me
when I was nine.

"¿Where are *you* headed to?"

"Down.
To Aconchi," I tell him.
Exactly 20 years ago the helicopter, the truck,
the detention cell . . .

I hear a teenage boy
being told he can't ask for asylum.

"There were others like him this morning."
The Mexican guard shows me his notes. "Look,
almost 257 just today,
all kids. We keep a paper record."

I stay inside the Mexican immigration office,
peer through the tinted glass
the teen can't look through.

He's alone. When I crossed,
I was parentless
but there were other adults.
No one is ever really alone,
I thought before,
when I could only read news
& not *be* anywhere near
the border without papers. But,
he *is* alone. ¿How?

Someone from Grupo Beta
tells the boy there's no migrant shelter in Naco.
The closest ones: Agua Prieta,
Hermosillo, Nogales.
¿Nothing?
"Nothing."
¿I can't stay here?
"You can't,
¿do you need a ride?
we can give you a ride," Grupo Beta says.

A la gran puta.
Dije que no iba a chillar. Estoy aquí.
Estoy allá. It does not matter. It
doesn't matter.

The teenage boy climbs into Grupo Beta's truck.
I want to help, but my feet won't move.
That was twenty years ago.

I couldn't stand in México
without being told I must
find a shelter. Without
being afraid of Mexican cops
& American B.P. agents. Since then
I've looked over my shoulder
for uniforms, always
ready to outrun, out-jump,
slide under wires, hide
in the brush. Because
some of the people with me
never made it, are in this dirt,
underground, & it's perhaps why
I've never had the urge
to touch the wall's rusted
metal slats. Draw on them.
Paint them. Even for this boy
who must be hungry, thirsty,
tired of being told he can't keep walking north,
who is driven away from the border
to spend yet another night
thinking tomorrow he'll cross—
I can't even bring myself
to protest it.

Javier Zamora

Javier Zamora was born in El Salvador and migrated to the US when he was nine to reunite with his parents. His first full-length poetry collection, *Unaccompanied* (Copper Canyon Press, 2017), explores how immigration and the US-funded Salvadoran civil war have impacted his family. Zamora was a 2018–2019 Radcliffe Fellow at Harvard University and holds fellowships from CantoMundo, Colgate University (Olive B. O'Connor), MacDowell, Macondo, the National Endowment for the Arts, Poetry Foundation (Ruth Lilly), Stanford University (Stegner), and Yaddo. He is the recipient of a 2017 Lannan Literary Fellowship, the 2017 Narrative Prize, and the 2016 Barnes & Noble Writer for Writers Award for his work in the Undocupoets Campaign. In Fall 2022, Hogarth will publish his memoir, *Solito.* He lives in Tucson, Arizona.

Alan Pelaez Lopez

A Future, Elsewhere (2020)

Alan Pelaez Lopez

Alan Pelaez Lopez is an AfroIndigenous poet, installation, and adornment artist from the coastal Zapotec community of Oaxaca, México. Their work is invested in thinking with and through experiences of fugitivity, the failure of language, grief, ancestral memories, and the role of storytelling in migrant households. Pelaez Lopez is the author of *Intergalactic Travels: poems from a fugitive alien* (The Operating System, 2020), a finalist for the 2020 International Latino Book Award, and *to love and mourn in the age of displacement* (Nomadic Press, 2020). They have been organizing with undocumented migrants since 2011, including cofounding Familia: Trans Queer Liberation Movement and the Black LGBTQIA+ Migrant Project.

lois-soto lane

Searching for Atlantis

When my father said we were immigrating to Atlanta, I thought he meant Atlantis. To an eight-year-old, it seemed full of magical promise and fun possibility. Atlanta was in America, the place the adults got excited about when they talked. It was made only sweeter by the prospect that it would just be me and my father. The constant heckling of aunts and relatives meant we'd never been left alone together. He wanted me to study in America. America had a better education system, he said. An American degree is unchallenged.

After my mother's death in childbirth, he had sent me to live with a distant aunt for some time, lamenting that I needed a woman's touch. When I was returned to him, my godmother took over raising me. All my father did was discipline me, and for that I grew to hate him. I had dreams of my real father coming to get me one day because it just seemed impossible that my father could love me and hurt me. And the discipline hurt. It was always a beating or being told to pick rice or kneel with my hands up and my eyes closed. The disciplining made us enemies. He towered above me like a giant, but there were also times when he allowed me to climb onto his lap and rub and slap his round bald head as it glistened in the light of the sun. In those moments he wasn't threatening. He seemed more like an overgrown bear that had wandered out from a forest years ago and never returned. I thought that once in the US, he and I could finally make a home together.

Atlanta was not Atlantis.

Having never lived anywhere below 96 degrees, the cold was all too new to me. Those first months were harsh. We spent the majority of our time trying to get me enrolled in private school, and we were constantly met with racism. I remember taking one entry test three times because the school administrators couldn't believe I was testing so high.

In our search for a school that would admit a little girl from Nigeria, we were met with a lot of questions. But the one that still hurts is when the principal of a charter school asked my father, *"And what about your wife, how dark is the mother?"*

If anything visibly changed in how my dad was holding himself, I didn't see it. But his voice was steel when he said, *"My wife is dead."*

My father did his best to shield me from the overt racism of the adults, but he couldn't do anything about the kids I went to school with who made fun of the way I talked and smelled, the ones who asked, *"Do you sleep in trees and eat monkeys?"*

When he asked me about school, I learned to lie, making up friends and stories that I wished were true. I didn't want him to feel he had to worry about me. I didn't want to bring him *wahala* and risk him regretting bringing me to America. I hadn't forgotten the people who'd told him to leave me behind. *What use is a girl in America?* I was diligent with my grades so he would be proud, devouring everything I could read and rehearsing my times tables under my breath. He liked to quiz me randomly, turning to me suddenly throughout the day and asking me to recite up to 13x. *"Your mom was good at math, she liked numbers,"* he told me once. *"She was an accountant."* That was the closest we ever came to talking about her.

Struggle was new to me.

In Nigeria we had been comfortable, what some might even call wealthy, and well-established within the community around us. We lived in a compound of aunties and uncles whose attention and hands I passed through comfortably. House girls cooked and cleaned and kept us out of mischief. There were always new gifts to rip into when my dad returned from a work trip, my favorite a life-size Barbie I took everywhere with me. Looking back, I realize that the cornerstone of our wealth came out of living in a community where resources and money easily exchanged hands. That compound of aunties and uncles acted as a safety net. Immigrating plunged us into anonymity and left us with nothing beneath our feet so that when my dad couldn't maintain the job he had back home, we scrambled to find family to lean on.

Family came in the shape of a new wife, my dad reverting back to his belief that I needed a woman's touch. I was excited at the idea of being a flower girl but angry at the new addition to our lives. I warred with myself over wanting a mom but wanting my dad to myself more. The marriage didn't last though. In those first months he was still living between Atlanta and Abuja, trying to keep his job as an oil trader for ExxonMobil. His marriage was less a union and more a means to an end. His ultimate goal was finding someone to mother me during his absences. Before, it had been nannies; now, it would be a wife. If I'm being honest, I can imagine that, in his eyes, they were more or less the same thing. His new wife wasn't prepared for an absentee husband, and they fought; her only ammo against him: me.

Between two adults a child can quickly become a casualty, and I found myself being used in fights in ways that still remain with me. We lived like this until my dad lost his job. Ironically, the fighting intensified until they went their separate ways. At eight I wanted to believe it was because my dad couldn't be with anyone but my mother, that she was his one true love. At twenty-three, I realize that it's because you can't build marriages on convenience.

Ever pointed in the direction of family and seeking the security and care of relatives, my dad moved us to Virginia where a cousin of his was building a life for himself. Virginia sounded like . . . Virginia. The luster of our adventure had worn off long before, once I realized it wasn't going to be the father-daughter extravaganza I had imagined.

In Virginia, my dad got a job as a home health aide and we settled into having less. Having less meant that when we talked to people back home, I lied and said, *Everything's fine!* Having less meant that I didn't do extracurriculars like the other kids at school. It meant there were no gifts, no new clothes, no trips. It meant I never learned to play an instrument, and I never went on planned school outings. Having less meant I didn't have friends over because we lived in a basement, and I slept on the couch.

The first time I had a friend over she said, "Now I know your secret. You don't have a room." It terrified me so much, I wanted to hurt her for seeing. And how did she see, I wondered, when I brought her into the most beautiful part of the house, the upstairs with the upholstered chairs, the brocaded curtains and lush carpeting and rugs. How did she see when I'd cleaned diligently in anticipation of her visit? *What* did she see?

The thing I never told anyone is that I was embarrassed—of our lack, of our constant needs. I was embarrassed of my father in a way a child should never learn to be. He had become small in the way that he carried himself. He went to work and came back less and less like a person, as if parts of his spirit died every time he labored. He was used to being the Big Dog, a C-level exec with an expense account and someone else to worry about his calendar, but that was no more. I was embarrassed about what other people would think. I was embarrassed by his words when he opened his mouth, thick with a Nigerian accent I no longer shared. Most of all I was embarrassed on his behalf because I felt his own shame at not being able to maintain the standard of living we were used to.

I was ashamed of myself and of my father and of the place that we came from. I learned this shame from other people much like I learned that I was Black. I learned this shame from the image of Nigeria that exists in the West, the image of the entirety of Africa as this backwards, uncivilized, "third-world" place. An image frozen in time with no context or grace.

The children at school made sure I never forgot any of these things. They took my name, my mother's name, the only thing I had left of her since she died and made it a joke. They called me Ebony, Ibonbuneh, Ibeanie, Ibienem. Anything that is not my name was what they called me, and I let them because I started to believe the lie.

I was ashamed of other things too.

I was ashamed I liked being with girls more than I liked being with boys. I was ashamed of being a motherless child. My father didn't talk about her much, and I now understand his reluctance. She belonged to him in a way she could never be mine—a real, breathing human being. To him she was not just a dream, but a ghost. I hated him for that too.

In some moments I also hated my mother. I felt othered by the loss of her. I didn't know how to talk about missing someone I had never known. I had no claim on her that felt tangible or real. She was my father's; she was never *ours.* It was as if I had no right to her. In my adulthood, though, I have staked a claim to her that I am delicately feeling my way around. I honor her on her birthday and keep a photo of her on my altar. She is mine. I talk openly and often about her, honoring the feelings whenever and however they present.

As a child, though, everything I felt for her—like our status as undocumented immigrants—became a secret. In those days our mouths were always full of secrets.

We're Black first before we're immigrants, and that, we quickly learned, is a killable offense. Our Blackness protected us from the threat of deportation—ICE policy and execution fixates on the southern border and Latinx immigrants—but it couldn't protect us from violence. For a long time, I didn't even know other undocumented Americans. That is the thing about living your life in secret: it isolates you. Outside of my father, I thought I was the only Black undocumented immigrant in America. That we were undocumented was my father's secret. He never shared with me anything about the politics of our being in America, and why should he have? I was a child. What I knew of borders and citizenship could fit through the eye of a needle.

Still, he fostered in me the dream of education. Told me to hold on to that because when you're smart, they can't touch you. *"You go far when you're smart,"* he said. The dream was that my intelligence would propel me all the way to college. *"Look at Obama."* But the dream didn't get very far; Obama had never been undocumented. My father hadn't considered that our being undocumented could stop me from achieving the dream. To him it wasn't even a thought. We came for a dream, and the dream we would get.

Our migration was hardest on my father. It had taken courage to leave the only land he had ever known, the land that joined him to his culture and made him a triumph. In America he wasn't even a man; he was something less than that. In my childhood ignorance I wasn't fully aware of what it meant for him to be a Black man in America. I didn't imagine that the bullying I faced as a result of the American Race Industrial Complex translated to adult relationships too.

If you look up my father's name, you will see a mug shot

from 2008 when he spent a night in jail just for being a Black man in Georgia.

My father was a prideful man, and all these humiliations took a toll on his health. Soon he was coughing all the time and chewing garlic like a desperate vampire. The doctors said he had diabetes.

That is the thing about America: it makes you physically ill.

The summer before I started high school, America made him so sick he had to go home to Nigeria. That's one version of the story he told me. Another version was that he was going home to be able to work for more money than he'd been able to make in America. For my college fund. The first mention of our status in the US came up subtly, in the Brown/Black way of speaking around something rather than getting at it directly.

"When I leave, I won't be able to come back."

No one said the word "undocumented."

Instead, the pressure that'd been building since we immigrated reached its peak. In his leaving, I represented not only myself but him. There was no discussion of whether I would be accompanying him home. Like all the decisions that were made in those years, what I wanted didn't matter. I don't think it occurred to him that I would want different things.

"*Be a good girl. Make good grades. Don't get into trouble. Stay focused on college. Don't embarrass me.*"

I wish I could say I cried when he left, but we didn't know each other well enough for the prospect of missing him to be sad. Somewhere in me, without my realizing it, I had already done the mourning for my father and said his last rites. I was thirteen, and we were still the same strangers we had been before we immigrated, perhaps even greater strangers, considering the secrets we kept from each other. America took the both of us and made us entirely new.

The ways in which I have missed him over the years are never enough to bridge the distances between us, but they do remind me of just how much he gave me. My father gave me my laugh. He was the one in the room always cracking jokes and setting up elaborate punch lines. He gave me my love for films. Some of my fondest memories of Nigeria are of the afternoons he would take me and my god siblings to the cinema and buy us all the popcorn and candy we wanted. Above all else, he gave me a chance to be more than I could've been in Nigeria.

I have lived in America longer than I ever did in Nigeria, and in the eyes of my countrypeople, I have been Americanized. And maybe that is true. I have become so much here: graduate, lover, writer. At twenty-three, I have built a small life for myself and found a community to fight the fight with. America hasn't been Atlantis—to be Black and an immigrant here is to be in a perpetual state of emergency—but I have found things worth treasuring and preserving while helping to fashion a new world for myself that brings me and my community closer to paradise.

lois-soto lane

---

lois-soto lane is a chronically ill and disabled transfemme Black lesbian immigrant anarchist writer and community worker by way of nigeria, living in mutable timelines. he is most concerned with the stories that take shape in the glow of indomitable truths and works to center the power of the quietest voice in his craft. through the grace of chosen family and the guidance of allah she is building a home of her own in brooklyn where she's working on an oral archival histories project focusing on the lives of Black undocumented immigrants and Black trans/nonbinary folk. their work has taken several shapes and can be found in several publications, including *Plantin Mag*, *The Unplug Collective*, *Crossin' Borders Magazine*, and *Hooligan Mag*. her work explores the realities of living in displacement, womanism as religion, dissonance in identity, & other bits she wants people to talk about more. she is rooting for everybody Black.

Jesús I. Valles

POEMS

Quinceañera

It was always your obsession with things that look soft and aren't. Tulle, crinolines, starched taffeta, the textures of all the things you've always thought most beautiful amidst the dirt.

It was always what starts in the dirt, your mother, Rosa, pinning clothes in the tendedero, white panties, the fading paisley of work paliacates, once-white A-shirts drying into ivory, the heavy, sogging, beige things to be unhooked after work, and worlds and worlds of sheets, blinding and flailing while the wind kicked up terregales en los barrancos de Ciudad Juárez.

Once, a man tried to steal those sheets. Having heard the rustling of his theft, the nervous foot's grind against the dirt, the rock's snitching under his soles, the silhouette of a man taking through the bathroom's foggy window, your mother left her shower, nearly naked, chasing the man with a brick, dirt and flowing sheets around her furious body, like a carapace, like a dress. An A-line marvel, swirling about her, cinched in right at the "¡Pinche viejo ratero!"

It was always the sand, enveloping the bones of the women who were bodies once, then teeth, then the faces of girls brutalized with staples onto a rotting utility pole, then front page covers on newsstands, the pages flowing too, hanging to the taut string by ridged mouths, the wire jaws of wooden clothespins, as if a body could ever hang on.

It was every princess fantasy you had as a boy, everything glimmering and flowing and rough was there, in that dirt. Everything tightened and gorgeous and boned into its silks. It was that picture of your sister, Yesica from her quinceañera; the wide-brimmed white hat and the gloves and how dumb this fifteen-year-old looked in a wide-brimmed white hat, and the peculiarity of gloves in Ciudad Juárez and how mesmerizing the gloves were, their opera length, the odd choice of ruched satin, the perfect shine of them. She helped raise you because your mother was cleaning other people's homes just a bridge away and you loved the world she brought when your mother went away.

Yesica's quince was a thing of dirt and ingenuity. Yesica at fifteen, beautiful. And rough. Like the tulle, and the wire, and the hoop skirts, and pins that come undone in the fitting; but she was like that always. The crew of jotos de la vecindad around her at all times. Your mother insisted she have five damas for her quince, but she refused to share the spotlight with anyone but her crew. For her birthday, they gifted her a move-by-move re-creation of Madonna's "Vogue" performance from the 1990 MTV Awards. You ever seen jotos dance in sand and wind? Yes. Divas in the dirt, serving French court on Chihuahuan soil. Yesica floating in that dress, every stitch, sequin, and bead, every blessed artifice, dazzling us away from city's cruelties.

It's an odd city. Fifteen meant maquilas. Fifteen meant fiestas. Fifteen meant bizcochos. Fifteen meant men drinking. Fifteen meant dancing late into the night. Fifteen meant bus rides to factories where the world was made. Fifteen meant everything about her small world was terrifying and sequined and breathing and choking all at once. Fifteen and the fear one day she'd be swallowed by that desert. But goddamn it, if she wasn't going to have her fucking quinceañera in Cd. Juárez. Goddamn it if she wasn't going to be fifteen and alive. Rough and soft and flowing and beautiful in the dirt—just like all the things you've always loved most.

Just a few weeks after her quinceañera, Yesica crosses the Río Bravo into El Paso. Her white gloves wilt yellow and rubbered, and there were Jafra perfumes and tardeadas in Dallas and cleaning other people's homes; she was alive and breathing and no sand ever ate her.

you find home / then you run

a poet says / she sees music in my work / "you build this speed / this motion / then you stop / then you go / it's like / you find home / then you run"

it's like that / like / i find myself next to you / want to leave you in the middle of the night / or like my mother loves me / but I can't fix her sick / so I get on a bus anyway / or like / every man that has wanted me into a harbor / then the land comes undone / or like / when i laid my body atop the first boy i ever loved / until the janitor walked in / the first thing i did was run / or like the way i seldom call home / but all i can talk about is my family—

is it like that?

a desert child / i want to look at water / long for a Pacific / pouring itself into the Rio Grande / and not mourn anything / is it like flowers / is it like roses / is it remembering every dirt like a limb / my own errant right leg / its left twin longing to root / or a house i once lived in / a home / then you run / like when Hector was my fifth-grade friend / the well-made promise of safe walks home / my trailer-park hero / beautiful then / and i was a round pigeon shaped thing / he / he / god / he was so fucking beautiful / later / when he had lupus / i would have turned into bone marrow / to live inside him / hold his body upright / is it like once / a man came to me / asked me to be inside him / he saw my painted nails / said / will you do this to me / i did / when his nails had dried / i told him / i have heels under my bed / I kissed between his legs / up along his feet / crowned them with black patent spikes / he asked for a dress / i slipped his body into the only gown i own / i birthed someone who emerged from the only womb i own / that person was beautiful / they left / after a long night / how beautiful / limbs emerging from tulle and nylon / stretch fabric / is it time?

for years / my mother would take / the clothes American women she cleaned for / gave her / hand-me-downs / their garbage / fill a wardrobe / for years / she would style / women looking to cross / the border through the bridge / she would dress them in white women's excess / style them / teach them how to say / "American" / to cross the bridge safely / American is a costume / my mother and i / perhaps / have so much more in common / I have a suitcase full of costumes / football jerseys / business suits / vinyl skirts / leather harnesses / I put on for Americans / I find a home in a suitcase / a red suitcase always ready / because nothing ever feels like home / like—

what if / the only part of my life that fits / in a passport photo / is that my mother loves me / and she will die one day / a series of passport photos / i always escape beautiful things / a chubby boy loved me / Philip / made me spam musubi / cried about his mother / her legs / i remember

worrying / next to his body / that she would die soon
/ because i would not know what to do if a boyfriend's
mother died / so maybe it will end soon / i will leave before
his mother dies / is my home my cowardice? / is it like
that? / or is it

that i love my friends / because they are the only country
i can ever know / México wanted me dead / this place
wants me jailed / then dead / like so many poets / long for
the desert / the cactus / the moon / like they have a home
there / i was born in the dirt / raised by a choir of factories
/ the lungs inside each factory / were all made of dead
women / women mourning their dead / who'd eventually
die / how do i write a beautiful sonnet / about the fact that
a maquiladora / made me / my mother ran from that place
/ my father stole me away / into this country / how do i tell
my lover / that i do not have a country

so it is / is it like that? / is it like i run from my friends /
who are my only home / like Ryan who holds my wild
in fistfuls / like cackling wild with Michelle over crooked
wigs / like Miss Taji's 2 a.m. serenades / like Eddie who's
got my funeral plans / O beloveds / they are / my only
country / i'll run away / as soon as they leave town /
because that is how it all happens / or that this city / is
always a room / with a bed / the man i loved most fled my
body / made a consulate of my mouth / my papers snagged
in his teeth / his papers held on my lips / like the last
morning / i saw him / he hugged me / "we should hang out

again" / never did / so i don't keep anything in my room / just the bed and suitcases / "you find home / then you run" / so then—

then you find home / then you run / then you run then you run then you run then you run / on.

then you.

Jesús I. Valles is a queer Mexican immigrant, educator, storyteller, and performer based in Austin, Texas, originally from Cd. Juarez, México. Jesús holds a master's degree in communication studies from California State University, Long Beach, with a focus on performance and qualitative research methods. Jesús is a recipient of the 2018 Undocupoets Fellowship, a 2018 Poetry Incubator fellow of the Poetry Foundation and Crescendo Literary, runner-up in the 2017 Button Poetry chapbook contest, and a finalist of the 2016 Write Bloody Publishing poetry contest. Their work has been published in the *Shade Journal, The Texas Review,* and *The New Republic.*

Danyeli Rodriguez Del Orbe

Pa' Nueva Yol

When I was four, Papi knocked on the window of my grandmother's house, where I lived with Mami.

"Ya me voy, Yeyi," he said, as he peeked through the blinds, motioning for Mami to open the metal gate wrapped around the veranda. Once inside, Papi picked me up and carried me into the living room where he sat me on his lap and declared that the day had finally arrived: He was leaving for Nueva Yol.

In San Francisco de Macoris, Dominican Republic, migration was common. "New York" and "the United States" were synonymous terms that signified a place better than here. Almost everyone in town had at least one loved one abroad. As a child, I began to know Nueva Yol as the place everyone was eager to go to, but rarely ever came back from. Already accepting this as Papi's fate, I began to cry into his shoulder.

"Por favor, no te vayas, Papi," I begged between sobs, until Mami had to detangle me from around his neck.

Papi had previously left on a yola, a medium-capacity, wooden boat used to cross the treacherous Caribbean Sea from the Dominican Republic to Puerto Rico via El Canal de la Mona. Once in Puerto Rico, Dominicans purchase travel documents to reach the US mainland. Hundreds of people a year die on the journey as yolas get lost at sea or flip in the middle of the ocean, leaving dozens at a time to drown or starve to death if not rescued in time. Over and over again, Papi was deported from Puerto Rico, but he'd now found a way to migrate via airplane directly from the Dominican Republic to the US.

Two years later, a week before my seventh birthday, Mami also left. Unable to financially support me on her own, she became increasingly exhausted and frustrated with her life on the island. Papi barely ever sent her money from abroad to cover my expenses, and when he did, it would not last long. Without much choice, Mami was forced to join the large number of single Dominican women migrating to the US. The morning Mami left, she kissed me while I slept, as she did every day before work. But later that same evening, she called from Tía Tina's apartment in Nueva Yol to let me know she had arrived safely at her new home. Mami knew how much I had cried and missed Papi when he left. So now I resented her for not telling me she was leaving. For needing to leave. For not taking me with her. Tía Mamin held me in her arms as I cried for Mami to return.

"She's going to send you lots of toys. You'll see," she whispered in my ear, but I was inconsolable.

Years later I asked Mami why she never said goodbye, and very matter-of-factly she answered, "I would not have been able to leave if you had asked me to stay."

After Mami and Papi left, I often played a game with my godbrother Fonsito. Whenever either of us spotted a fallen eyelash on the other's face, we would close our eyes and make a wish. *"Irme pa' Nueva Yol,"* I would repeat in my head, as I envisioned playing in the snow like those white kids in American movies. The soft, white slush turning my fingers numb like frío-fríos did. The tall buildings decorating the sky with lights brighter than the stars. The smell that escaped Tía Tina's luggage when she distributed gifts during her visits from abroad—a blouse for Mamá to go to church, a pair of jeans for Tía Damaris, an outfit for Mami, new sandals for me. *"Irme pa' Nueva Yol, irme pa' Nueva Yol,"* leave for New York, I mouthed over and over again.

When I was eight, all those wishes se cumplieron. About a month before I boarded a plane to New York, Mami explained on the phone that I needed to start seeing a woman who would be responsible for bringing me. A few times a week, I would visit the woman's house, where she would help me memorize a new name, parents' names, date of birth, some US history, and a list of phrases often used in the US. I did not understand, until I was much older, that I had used someone else's papers to migrate. The terms and history I was to memorize—for example, What do the fifty stars on the US flag represent?—were useful for me to know in case Customs and Border Patrol agents questioned me. In September 2003, I finally reunited with Mami.

Mami had three jobs: waitressing six days a week in a Dominican restaurant between Washington Heights and Harlem, bartending on the weekends, and cleaning mansions in

Eastchester once a week. Her lack of English and college education kept her from higher paying jobs. For a few months, Mami and I saw each other in passing before she left home in the morning and late afternoon when she came home to change into her bartending clothes. I spent a few hours alone after school, until my cousins, godmother, or aunt came home from work. For so long I had dreamed about reuniting with Mami, only to find that the reality of migration did not allow for much time together.

Things were no better at school. Two weeks after arriving in the US, Mami had enrolled me in P.S. 15, where I entered the fifth grade. As a gifted learner, I had skipped two grades in the Dominican Republic. The age difference between my US classmates and me, made worse by my lack of English, fostered an environment of cruelty at school. Once, a classmate choked me while pinning me against the wall for mistakenly cutting in front of her in the line to the lunchroom. This kind of physical, mental, and emotional abuse continued for the rest of the school year until I was eventually forced to enroll in another middle school. My school days were excruciatingly long, and I often came home begging to be sent back to the DR, to the comfort of my grandmother's house. The eagerness to live in Nueva Yol quickly wore off and transformed into a longing to return.

My relationship with Papi in the US couldn't have been more different than how it was on the island. When a family friend told him that Mami had brought me to Nueva Yol, he showed up at Tía Tina's house filled with rage, convinced that I was better off on the island. He threatened to take me with him and call immigration officials to have Mami deported. Their relationship became increasingly tense throughout the years, and Papi stopped being the attentive, loving father he used to be. He came in and out of my life, calling every few weeks or dropping by every few months. Watching how Mami's health declined from working so much while Papi refused to provide any financial support, I became angry and resentful. By the time I was a teenager, we were unable to hold a conversation without erupting into explosive arguments about his shortcomings.

In the thick of survival, Mami and I loosened the grips on our country, tucking away our hopes for return anytime soon. Mami took English classes at Bronx Community College, and I focused on unrolling my *r*'s and burying El Cibao in a casket of proper Spanish. Mami and Tía Tina ironed the countryside out of my tongue, making sure I no longer replaced my *r*'s with *i*'s like campesinos on the island did. *Poique* became *porque* and native Dominican words like *guagua* became *el bus*. With time, my Spanish underwent yet another transformation when I gained enough courage to speak more English.

By the end of high school, I was more Dominican-York, more Dominican-American than I could bring myself to admit. I stopped watching Dominican TV channels and telenovelas and watched Disney Channel and rom-coms instead. Vanessa, my best friend, and I used to come home from school and sing karaoke to pop and alternative rock songs until our voices were hoarse. I thought that by forcing myself to like "white music" I could become more American. The closer to whiteness I felt, the less I thought it mattered that I was undocumented. Naively, I thought that legislators would one day hear about my assimilation and affirm: "You're a real American. You deserve legal status."

However, being undocumented and knowing I would not return home anytime soon taught me to appreciate the limited ways that New York offered me access to my native country. New York, home to the largest Dominican population outside of the island, was special for many reasons. It boasted a small movie theater on the corner of 181st and Broadway that showcased Dominican movies. Whenever we went to see a screening, Mami and I experienced a kind of togetherness, a sense of normalcy and belonging that we had experienced only back on the island.

On weekends, Mami treated me to habichuela con dulce, a Dominican dessert often cooked during Lent, that a Dominican woman sold out of a small food cart on the corner of 181st and St. Nicholas. We ate authentic Dominican food at El Malecón and Caridad restaurants, in uptown Washington Heights. Dining among Dominicans, hearing the cooks yell out orders from somewhere in the kitchen in Cibao slang, observing as Dominican servers talked with their hands, and hearing the clinking of dishes dancing in the air with whatever merengue or bachata song played from large speakers showered me in comfort and familiarity and eased the pain of nostalgia and longing I felt for my native country.

In adulthood, I settled for Dyckman in Inwood, Manhattan, the strip of bars that offered Mami her first bartending job. During the day, most Dyckman establishments are family restaurants, where loud music interrupts any attempt at conversation. But after 10 p.m., restaurants become hip clubs, where DJs play dembow, bachata, merengue, salsa, hip hop, and amp up the crowd by shouting, "¿Dónde están las mujeres que no tienen mari'o?" over every other song. These clubs are undoubtedly Caribbean, specifically Dominican and Puerto Rican. In Dyckman, I could touch my country by putting my hand on the shoulder of any man swaying his hips to las tamboras. I could smell the salt in the air as if I were driving, windows down, by the beach in Nagua. With every dembow, Aventura throwback, and Toño Rosario song, I could walk to my home across the ocean, heels in hand, and climb into my grandmother's bed in the middle of the night as I used to when I was six years old.

For a while, I pretended that this was enough. I had to. I had already spent most of my childhood and adolescence aching for home and cursing my fate. To ask more out of life was to set myself up for disappointment. Mami and I had waited years for "una ley," any legislation that created a path toward legalization, but all we ever got was politicians arguing over our livelihood on the 5 p.m. news.

Around 2013, a decade after migrating, an attorney whom I interned for screened me for immigration relief. By then, things with Papi had deteriorated, and she determined that my on-and-off relationship with him made me eligible for Special Juvenile Immigration Status (SJIS). After I had had no contact with him for more than two years and no financial support for more than four, the courts determined that Papi had abandoned me, therefore fulfilling the requirement for

SJIS. The morning of my college graduation, U.S. Citizen and Immigration Services interviewed me and approved my legal permanent status.

This is how, a week after my twenty-first birthday, I boarded a plane alone en route to the Dominican Republic, more than a decade since the first and only other time I had been on a plane. Once in my seat, I opened my book bag and took out a copy of *Questions for Ada* by Ijeoma Umebinyuo and found the bookmarked page where I had placed a sticky note and written *"for the return home"* years before. I marveled at her words: "i will return home to five graves / to ancestors who held me as a baby / telling me who i was once in my former life / never a foreigner / always the daughter of her people," and I released the longing I had held tightly in my chest for twelve years.

Danyeli Rodriguez Del Orbe is a Dominican, Bronx-raised community organizer, writer, and spoken word performer. Her work raises awareness around issues of race, gender, and migration. She has been featured by the Bronx Museum of the Arts, the San Francisco Museum of the African Diaspora, *People en Español* magazine, and others. In December 2019, she self-published her first collection of poetry, *periódicos de ayer, a lover's archive.* Danyeli is also the cofounder and cohost of *loose accents*, a Latinx podcast that highlights the immigrant experiences of the East and West coasts. In addition to performing, Danyeli has been an avid advocate for immigrant rights, receiving the New York State Dream Activist Award and being a recipient of an Immigrant Justice Corps Justice Fellowship and a New York Immigration Coalition DREAM Fellowship.

Rita E. Urquijo-Ruiz

First Visit

—Mom cries for you almost every night since you left, your oldest sister says.

You often fight the urge to cry too, especially on Sunday afternoons when your family in México gets together and life goes on without you. It's been five years since you crossed the border to chase the dream of learning English and then working as a bilingual secretary in your hometown. You were sixteen then. Now you have bigger goals. It's spring 1992 and you're a sophomore at the University of California, Riverside. You never thought you would make it this far. And you don't want to risk it. But there's a longing inside you that grows by the day—to see your loved ones, especially your mom. She'd been raising your ten siblings by herself since your dad died, but with a third-grade education, she doesn't make enough money to support everyone. Your dream is to graduate from UCR, help out the family, and maybe even spare your mom the fate of being a maid into her old age.

You've wanted to go to Mexico for some time now but getting caught at the border on the way back, being deported, and losing out on a bachelor's degree from a US university is too big a risk. Not to mention jeopardizing your chance to legalize your status.

Mr. Porter, your math teacher at Artesia High, in Lakewood, California, was so moved by your undocumented situation that he and his family took you in for your junior and senior years. With their support and thousands of dollars in lawyers' fees, you began a seven-year process of getting your papers.

But you know you can't wait that long. Something moves you to go see your family back in Sonora. Yet, how would you come back?

Your UCR friend Alex says—Go for it! Just tell the migra at the border that you're a US citizen. In your two-year language immersion with the Porters, somehow you picked up their southern California accent, and your friends tease you that you sound like a gringa when you speak English. You let out a big sigh, smile, and say, "Fuck it, I'm going to México!" And so, you do.

You're nervous when the Gonzálezes, the first family you stayed with when you arrived, pick you up in their van at your campus apartment where you have your own big room with a door that locks. Long gone are your childhood days of sharing one doorless bedroom with ten other people. You say "hi" to everyone, get your luggage into the van, and squeeze somewhere in the middle. You're talkative and fidgety because you're excited and nervous heading home. You don't pay much attention to the eleven-hour drive except when the old, beautiful saguaros of the Sonoran Desert start flanking the lone road—the first to welcome you home.

When you get there that Saturday, everyone in your barrio comes to see you. Irene, your childhood best friend, surprises you with a beautiful floral arrangement. Receiving roses is rare in your neighborhood, especially if it's not for a funeral. You go get your camera to capture the moment, but your Tía Rita stops you.

—Mijita, get a clean sheet and hang it on the clothesline as a background.

She wants you to cover the unsightly yard—discarded furniture, old tin and aluminum cans, piles of cardboard ready for recycling, and trash to be collected soon. As you imagine showing the pictures to your friends in California, you're relieved by your tía's idea and run to get the bed sheet and help hang it. Five years of paved roads and fancy houses with manicured front and back yards have offered you much comfort. You chastise yourself for being "too gringa." You swore not to complain about anything that might embarrass your family.

About thirty people—family, friends, and neighbors—pose for pictures with you in your 1990s hairdo and big glasses, attempting to stay in the closet a little longer when asked if you have a boyfriend. (Years later you'll notice that in one of the pictures your mom is smiling proudly and holding you tightly by your waist. You'll indulge in this tender moment and recall it when you miss her the most. If you close your eyes, you can still feel your mom's strong, firm hand holding you close; you can even smell the Avon perfume she wore that day to celebrate your arrival.)

During lunch, everyone sits outside on chairs borrowed from the neighbors, to hear about your adventures "en el otro lado." Your cousin Blanca asks you to say something in

English, but you don't indulge her to not "show-off," even if it hurts her feelings. You'll make it up to her later by giving her a California beach-and-sunset key chain.

Your Nana welcomes you with her bendición, allows you to enjoy yourself and chat with everyone, but after eating, you know you'll have to face her. Finally, you overhear her saying,—Tell Rita Elena I want to talk to her, and your stomach tenses up as you head her way. Her adobe kitchen is smaller than you remember, crowded with trinkets and bagged dry and canned food on every flat surface. One of your favorite things on her table is her chipped pewter azucarera with three pink faded roses on it. Its two little handles remind you of your Nana when she puts her hands on her hips before scolding someone; fortunately, this someone was never you. She's waiting by the old table covered with a blue, floral oilcloth. Her saints and virgin statues stand tall in an altar to her right.

—How was your trip, mijita?

—It was fine, Nana. It went by fast.

—You are blessed to have the Gonzálezes and the Porters in your life. I am so grateful to them.

You agree with your head, smile nervously, as you anticipate the harder questions. Tracing the big yellow flowers on the oilcloth with your index finger, you look up at your Nana a few times.

—And do you have a boyfriend, mija? I know you're old enough, but you must be careful because men only want one thing.

—Ay, Nana! Don't say that. You know I wanna finish school first.

Your cheeks get hot, and you look down at the image of the flowers in front of you.

—No, I don't have a boyfriend now, but I've a picture of my high school ex and me at my senior prom.

As you reach into your bag to choose among copies of the photo, she's already shaking her head. Reluctantly, she grabs the 5-by-7 of you and your ex. He wears a tuxedo with a shiny, light pink cummerbund and bow tie that matches your borrowed taffeta dress. You tried hard to fit in with your straight friends. You even let them do your makeup and stiffen your hair up with Aqua Net. (You wished you'd been the one wearing that bow tie, and so years later, bow ties will become part of your signature look when you present your academic research at conferences and other public events.) When you tell your Nana that this boy is no longer in your life, she refuses the picture.

—No, gracias, mijita. Give them to your sisters and friends.

Your decoy boyfriend failed. You wish you could tell her about Pilar instead—your girlfriend back in Riverside.

Your Nana says she's worried about your papers, so you tell her that your best friend, Lalo, offered to marry you if the Porters' legalization plan takes too long.

—That would be good! If you get married, especially in a church, I want a copy of that picture. I'd feel so much better if you have a man to take care of you. I could finally die in peace.

You don't argue with her about her old-fashioned machismo. Not even your "La Chicana" studies class prepared you to persuade your grandma of how capable you are of taking care of yourself. You touch the watch Pilar gave you and think of her nervously, but smile anyway as you get ready to go back outside.

Your Nana gives you a laminated stamp of "La Sombra de San Pedro" to keep you safe on the way back and always. That day, as you walk away from her, you wonder what she would say or pray for if she knew about Pilar.

Each day seems too short, and after visiting with at least sixty people, mostly family and friends, on Friday afternoon, you finally have your mom all to yourself. You hadn't been able to see much of her. Her middle-class bosses said they were happy for your visit but couldn't pay her that week unless she worked. Your mom is smoking and sitting at her kitchen table preparing dinner. Pinto beans are cooking on the first gas stove she bought herself when your dad died. Even after a cheap paint job, you can notice traces of the soot in the corner where the old woodstove was. The aroma of the beans fills the room. The light outside is bright, but the little kitchen window doesn't let much in. You can hear the usual barrio noises—neighbors loudly talking, cumbias and norteñas at odds with each other, children playing on the street—it takes you back to your childhood. You remember Concha, the butch lesbian whom everyone called "marimacha," and how you'd stop and stare at her. She'd secretly wink and smile at you when she walked by wearing her tight jeans, brown boots, vaquero shirt, and hat. You'd run away scared yet intrigued as your friends and siblings bullied you.

By now, your mom is almost done with the prep work and looks at you.

—Elena, how are things in "el otro lado"? They treat you okay? Tell me the truth, are you happy?

She's the only one who calls you by your middle name; especially when she's being tender.

Sitting at the table, with some hesitation, you look at her. You're not as nervous as when you talked to your grandma but still feel like you cannot tell her the truth.

—Everyone treats me well, Amá!

You swallow the knot in your throat, delivering your best performance because you don't want her to worry. You withhold the countless moments when you made mistakes and feared you'd be sent back. You reassure her that you both made the right decision when she encouraged you to pursue your dreams.

—You sure?

—Sí, Amá, I swear.

—I'm so glad, mijita. And do you have a boyfriend?

—No, Amá. I don't want a boyfriend. I'm all about school.

You try hard to contain a smile as you think of your Pilar again and the fleeting possibility of coming out to your mom.

—Good, mijita!

She doesn't look at you as she chops some onion. After a moment, you feel brave and say:

—Amá, you know what? I don't think I ever want to marry a man. Maybe I'll have a child or two but I don't want to get married.

Without hesitation, your mom looks at you:

—Well, as long as you finish your degree, Elena you'll never need to marry anybody. You'll be able to take care of yourself, and the more you study, the better off you'll be to have the life you want.

She fries some garlic and onions, and the aroma promises a delicious dish of Mexican rice, beans, and enchiladas sonorenses. You smile because to you, your mother just gave you her blessing to be you, which, in your mind, means to be queer. You also know that this is probably the best you can do to come out to her.

Sunday morning arrives too soon. You must return with the Gonzálezes to the San Diego border. Before getting into the van, you stand in front of your mom and your Nana, and they each draw an invisible sign of the cross on your forehead. They have packed tacos and big, Sonora-style flour tortillas made especially for the trip north. Your Nana gives you an additional santito, a laminated copy of San Peregrino to carry in your backpack for a safe return. (Decades later, you'll transfer these saints to each butch leather wallet you'll carry in your pants' back pocket.)

You'd called your UCR friends on Wednesday to make sure everything was as planned. The van will drop you off a few blocks from the border crossing, and you'll take a taxi to Calle Revolución where Kuka, Teto, and Alex will be waiting for you.

When you arrive, they set out to get you drunk because they know what a chicken you are, and they don't want to get in trouble. Kuka has her green card ready and is going over her story about visiting her relatives in Guanajuato. She's the best actress in your Teatro Quinto Sol group, so she can pull this off perfectly. After two straight tequila shots and a tequila popper, you're ready to go. Kuka and Alex hold your arms, call you "light weight," and laugh at your drunken, silly jokes on your way to the parking lot. When you get in the car and sit next to Kuka, you start fidgeting and become quiet. You begin sweating and it's hard to breathe. She holds your hand, squeezes gently. It helps. Teto is driving and Alex is also in the front. Kuka sits behind the driver's seat for the migra officer to see her clearly. The line of cars feels endless. When there's only one car in front, your heart accelerates, and you feel sober in spite of all the tequila in your system. Your face feels hot as the gringo migra officer asks:

—Country of citizenship?

Teto, Alex, and you take turns saying "U.S.," and Kuka comes through beautifully. She's in charge of answering all the questions after she says "México" and gives her green card to the officer. He steps into his booth, types up some stuff, gives Kuka back her card, and waves your car through. After rolling up the windows and driving slowly over the speed bumps, all of you take a deep breath, laugh, and high-five each other. Interstate 5 is busy as you leave San Ysidro. You'll be in Riverside in the evening.

Shortly after your first visit home, your mom will be diagnosed with lung cancer that will take her away in six months. But for now, you feel that your family's prayers worked, and your dreams are still within reach. Pilar will be happy to see you, and you'll embrace her for a long time as you rest in her arms tonight.

Rita E. Urquijo-Ruiz is a Mexicana/Chicana queer educator, writer, activist, and performer born in Sonora, México, and raised in southern California. Her academic interests are Mexican, Chicana/o/x, and Latina/o/x literatures, cultures, gender, and sexuality, as well as theater and performance studies. As a child of the México/US borderlands, her approach to teaching and writing is interdisciplinary by nature, and her work centers the stories of socially and economically marginalized communities in both countries. She is the first member of her extended family to receive a college degree, and in 2019, she became the first Chicana/Latina faculty member to go through the ranks and become a full professor at Trinity University in San Antonio, Texas. She received Trinity's award for distinguished university, community, and professional service for her local and national contributions. She is the inaugural director of Global Latinx Studies and the codirector of the Latinx Leadership Initiative.

Kaveh Bassiri

POEMS

Caravan

With its "unknown Middle Easterners,"
Its Persian roots,
This karwan has been a long time coming.
It has traveled through deserts patrolled by time
Holding to the humps of Sanskrit saddles
Kidnapped by the crusades and footnotes.
From where the sun lifts its helmet of light,
There are grains of sands trapped inside the mouth of a boat,
On lorries, trains, and ferries,
Pilgrims packed together as armor,
Sweat dropping like shells.

They are coming to tell their stories,
Of nothing homey in homicide,
Of the long shadows cast by Lincoln.
Working night shifts on the back of La Bestia
Through fields the color of piss,
Racing with coyotes and clouds,
Crossing villages where mothers are waiting

SOMEWHERE WE ARE HUMAN

With bags of rice and beans,
Carrying flags as washcloths,
To clear the tables,
To harvest avocados, berries, textbooks
As guests of mosquitoes and stars
Soaking by the strands of the noontide,
Sleeping on asphalt shores,
Seizing the arms of their pedars and madars,
Savoring your dreams.

They are coming to build the walls and software,
Sheltering under the passport-colored sky,
Standing by the side of the road
In checkout lines at grocery stores,
With pupils deep as oil wells:
Ripe faces bearing the seeds of the streets.

Everyone awaiting: the families, the cops, the thieves.
Cell phones circle like flies.
Cells are their waiting rooms and caravanserai:
Serai, a Persian palace, the protector.

They are coming as refugees, resident aliens, dreamers,
Leafing out of the undocumented past
To translate themselves,
Lungs chiming in the body's camper,
Burying their words with the compost,
Past the meters keeping time on sidewalks,
Marching as a pack of matches
Striking and striking the pavement.

Kaveh Bassiri

They are coming with or without papers
To rewrite their stories,
To forget the anthem of the scarlet macaw,
To eat Buffalo Wild Wings,
To translate us.

Their footprints are like empty wombs,
History like stains from street lamps
To let you know you're not alone.
Behind them, the river of plastic bottles
And Styrofoam plates.
Behind them, empty shelves,
Stampeding dreams
Like scattered wavelengths of light.

Immigrant Song

We tracked the migrant sun that carts the day and ask,
collect our questions, take them to the savior's home.

A song, not for the moon that leaves Tehran each dawn;
tomorrow she'll amend and come ashore, home.

We found The Gap and Borders. Targets everywhere.
Why am I still complaining, bargaining for home?

And now I live with cell phones, microwaves, laptops,
Yet still. It's dust and leaves, I bring as mentors home.

How rich—secure—are dust and clouds that give themselves.
So free they live without a need for fence or home.

The subway ads bid sixteen million eyes to watch.
Is this how we are planning to insure our homes?

Contrition fills my cup, emptied for you, not love
or wine. It pours, for I have neither you nor home.

My toes and fingers, hairy quills, I dip inside
my bed to scribe the dreams about the Lessor's home.

The rooms you furnish with the foreign nouns, Gharib,
are lost, if none remembers, dwells inside your home.

Kaveh Bassiri

Learning Drills

Before the revolution was a street,
a holiday, before the pigments of war
were washed off the silenced bedrooms,
you could hear the footsteps of the pages,
follow along the black and white sidewalks.
The daylight still changed channels, radio
captioned hours, brought the evening's tinder.

It was a time when Lazy Bones' spine
was replaced by the beam of a flashlight
and the Space Commanders telegrammed
coded messages.
Grandmother tells us
that and further back to the times of jinn,
of dots so far apart you traced their bond
in sky, and blades so near you taste their sweat.

As the day leaves the backyards and night
reclines on the lawns, rooms burn like fireflies.
The divan, open like a blank book, holds
us together, as we survey the stars,
listen to our electric gun, monitor
with a remote, the night's chapters of Escape
Your Shape, Total Perfection, Creating Wealth.

Kaveh Bassiri is an Iranian-American writer and translator. He is the author of two chapbooks: *99 Names of Exile* (2019), winner of the Anzaldúa Poetry Prize, and *Elementary English* (2020), winner of the Rick Campbell Chapbook Prize. His poetry has been published in *Best American Poetry 2020, Best New Poets 2020, Virginia Quarterly Review, Copper Nickel, Beloit Poetry Journal, The Cincinnati Review,* and *Shenandoah.* He is the recipient of a 2019 translation fellowship from the National Endowment for the Arts. His translations have appeared in *The Common, Blackbird, Chicago Review, Denver Quarterly, Colorado Review,* and *The Massachusetts Review.*

Azul Uribe

10

My ten-year, post-deportation ban has expired. I've been outside of the United States for 4,110 days, more than a quarter of my life. My parents and siblings stayed in Texas to water all the bluebonnets for me.

Everyone always asks how I was the only one in my family who got deported. US immigration laws are without rhyme or reason. They're at once a game of chance, blessings, and circumstances, or in the case of the very wealthy, entrance fees. In my case, a babysitting side job gone wrong that resulted in a misdemeanor charge. After a grueling three years of fighting my case in court, I lost and was forced to sign a voluntary departure order, concurrently banned from reentering the US for ten years.

Coming-to-America stories are all alike. Returning-to-the-Homeland is different; no one does so in the same way. You leave or return, self-deport or are deported, each action laying a different foundation for life ahead.

After being deported, I tried to make a home in Hermosillo and Cancún until settling in Mérida in the fourth year of my ban. My mattress lay on the bare floor until the seventh year when I bought a bed frame. I didn't buy a Christmas tree until my ninth. In the US, I might be expected to have a college degree, a husband, children, and a pay raise by now. Instead, I have a cat, a boyfriend, three pairs of flip-flops, and two pairs of dress sandals. I've also learned to settle for the few meager employment possibilities in México for returnees who speak perfect English, the salary insufficient to cover basic living costs: as a customer service representative for stateside companies that subcontract with Mexican companies; as an English teacher for one of the many private language schools; or as a medical interpreter for a Mexican company that pays next to nothing, includes no benefits to its employees, people too tired and traumatized to push back.

I have not moved on in the traditional sense. The things I have here won't qualify as success, but I've learned to be content.

One day while running errands, I see a church I've never noticed before. A rule of thumb of my new life includes running a mental checklist of possible pitfalls, a quick and clean defense mechanism. Will there be new people? Meeting new people means I have to explain how I ended up in the Yucatán Peninsula. Acting on faith, I take the risk of attending a Sunday service. Father Jose's sermon is alive with Jesus, Karl Marx, and Judith Butler. He says, "It's not important that you believe in God. What difference does it make to an all-powerful God if you believe in him? What's most important is that you understand that God believes in you and loves you." His eyes are sparkling, and I can feel his words settling over me, wrapping me like a blanket. God is a warmth I can understand.

There's an accent to his voice and I later find out he's from Portugal; Father Jose is an immigrant, too. He is a psychologist. I remember my several failed attempts with psychiatrists and therapists here in México, specialists who, while well-meaning, gave advice like, "Don't be depressed about being deported, Skype and Facebook exist!" When Father Jose quotes James Baldwin in another sermon, saying—"It is easy to proclaim all souls equal in the sight of God; it is hard to make men equal on earth in the sight of men"—I muster the courage and make an appointment.

I spend my first session reminding myself to breathe because I'm overflowing, so much is coming out of me. Father Jose mostly listens to my fast, nervous sentences. I mention that the end of my ten-year ban from the US is near, how every time I try to look at it directly, it vanishes. I beat around the bush about feeling angry, saying how I am not angry, not really, it's complicated, because I have made mistakes, and what right do I have to be angry at the results? He interjects once with, "Oh, that's a holy fire." Father Jose is patient, his voice is level—"There are a lot of prophets who did God's work from a place of holy anger. It is sacred. Your anger is a gift"—and my insides feel like a room that's had its windows opened after decades of being closed. I can breathe. Once we say goodbye, I'm alone, repeating what he said to me. Anger is a holy fire. Anger is a gift.

The biggest success in building my current life—with contentment as the focus—centers on the friendships that have blossomed and survived as I have gone from devoted Mormon deportee, to abandoned-by-the-church returnee, and finally to Episcopalian with a question mark. After ten years, I can finally talk about being deported, sometimes without crying.

Abinadi, Vanessa, and Saul are some of the first friends I made in Mérida, all Spanish speakers, all Mexican-born, raised, planted, all living their entire lives within the limits of this side of the border. Our friendship is inching closer to the decade mark—nine years! It's a countdown that makes me happy. We go out for a drive and catch up on our lives. During a quiet moment, Abinadi's voice interjects with "Ya casi se acaban tus diez años, ¿no?" in the middle of discussing Mormon politics.

Deportation is not universally understood, and even among those of us who have suffered it, it's difficult to speak about. I tell them what happened, how I was charged with misdemeanor child abuse after babysitting ten kids. One of them ran out into the desert. The details of that afternoon are blurry: I chased after her, she swung a 2-by-4 at my head. I tried to hold her down, she slammed her head against my chest. At some point she called me a sweaty pig and wished me a heart attack. She pinched my thighs and arms so hard that my skin felt like ruin, with a secret language of paper cuts. I slapped her. Her parents came home after the fact to see me vomiting from anxiety and shaking, and they laughed. "She just gets that way sometimes," they said, and "Oh, we should have told you!" and "It's no big deal, really." At the time I thought I was slow to get the joke, everyone was laughing, and I needed to lighten up.

A week after the incident, the parents gave me the card of a police detective; *he just wants to talk*, they said, though later I would learn they were shining a spotlight on me to distract the police from looking too closely at them and their own negligence. I was led to believe that the meeting was just a conversation. The family seemed so casual about it, the officer even told me over the phone that it'd be just a quick chat, that I would be in and out after ten minutes. Mormonism has this ingrained, fundamental cultural notion to trust authority, and since the detective was Mormon, and the family was Mormon, I had no reason to doubt them even though I was terrified.

A few days later, I went to meet with the detective, and within minutes he booked me, handing me over to another officer to be strip-searched in the process. I'd never been seen naked by anyone who was not a woman in my family—I was saving that for my wedding day. While I struggled with feeling dirty and monstrous, the detective called ICE, and I was driven from the Iron County jail in Cedar City, Utah, to the immigration detention center located in St. George, forty-five minutes away. The ICE agents driving the van listened to Rush Limbaugh while the handcuffs on my hands and feet rattled in the back seat. It was my twenty-second birthday. I could see the temple garment line under their shirts, the telltale sign of Mormon attire, and thought how on any other day they might have called me Sister Uribe. Instead, they didn't even address me as a person when I got in, or during the drive, or when I got down in St. George, to be registered and held at the immigration detention center.

My deportation proceedings and eventual exodus took place in Utah, a predominantly Mormon state. During the three years my court hearings lasted, I ferried back and forth between whatever city in Utah I was living in and the Immigration Court, located in a strip mall in Salt Lake City, next to a Wells Fargo branch. Meanwhile, the peak of the Proposition 8 push was unfolding—a law in California that would grant marriage rights to same-sex couples—a fact that would bring itself to the forefront of my mind every time I heard a church leader say that the church was not political, and yet the church was rallying its troops to oppose Prop 8. They had never pushed back against anti-immigrant laws. I was wounded by God's lack of concern for my situation. Why was the immigrant cause not a cause for church action?

The car is silent for a moment, and for the first time, I don't cry.

Father Jose's house features a room with black bookcases that line all four walls; they stand twelve feet tall and still don't fully reach the ceiling. It's my favorite room in his house.

It's my second session, and I get right to the heart of my current anxiety: I am convinced I'm somehow dying, and I don't know how. I've had a cough that's lasted two weeks with a diagnosis of heartburn and a treatment of codeine syrup, omeprazole, and a special rinse to gargle. Father Jose listens to me tell him that I feel like I am dying, that it might be related to my ban. I've thought of my ban every day for the past decade of my life.

I tell him how when you're a kid and the dentist takes a tooth out, you run your tongue through the gaping hole where your tooth once was, even though the dentist told you not to, and the memory of that tooth insinuates itself with every sip or carelessly chewed bite. My ban is that hole, I say to him, and now I won't have that hole to run my tongue through.

In the time I've been gone, my little sister has gotten married and divorced, and my brother has had three more children. I've witnessed these events through a phone screen, a laptop, or photographs, each time punctuated with a "You were the only one missing!" and wrapped in a "We missed you so much!"

My niece was three when I was forced to leave. She liked Dora the Explorer and pretending to be a puppy, and she waved to me smiling, her tiny fingers gesturing goodbye. Now she is fourteen, reads as much as I do, and organizes her things for fun. I have missed a decade of her life. I missed her brother—my nephew's—entire life. He died a few years ago. I was Skyped into his funeral, the little casket a high-definition set of pixels. My niece was ten then, and she enunciated my name clearly, as she carried me around on an iPad screen, telling people, "It's my Aunt Azul," as if trying to explain the obscenity of a bright screen in a cemetery on an overcast day.

I tell Father Jose about these things, how the ban will be

up but those things will haunt me, a decade of holes for me to run through or walk into, shadows where I should have been. It is not a life. I don't feel suicidal, but surely, I must be dying.

It is a type of death, he tells me. Some things must die and from those we are reborn again, but we have to figure out what parts are dying and what we are being reborn to. The room full of twelve-foot bookshelves is a womb, and I am here gestating. I hadn't thought of any of these things, of other deaths and new lives. I hadn't considered I could be born again, or that some other part of me that wasn't my body was dying.

We can figure this out, he says.

Father Jose's explanations are stories, cousins to parables. He tells me, You know, there's a man who was born in the Isla de Azores, his name was Manuel and when he died, they put some sand in one hand and some dirt in the other, so he could have with him a piece of his home that he loved so much. When he arrived at the pearly gates naked, Saint Peter told him, "Oh, we are glad you are here! But you have to let that sand go in order to come in." And he said, "I can't, it is from Azores, my beautiful island home where my family and my ancestors come from." And so the saint says, Let me see what I can do, and goes in to talk to Jesus. Saint Peter tells Jesus that Manuel is here, but that Manuel insists that he can't come in without the sand from the Azores. "Let me see what I can do, let me go talk to someone," says Jesus, and he goes and talks with Gabriel, but really just plucks a feather from his wings and gives it to Saint Peter. When Saint Peter is in front of Manuel, he tickles the man with the feather, and because

Manuel is very ticklish, he lets go of the sand and finally the gates of heaven open up to him. And do you know what's inside? All of the Isla de Azores.

Father Jose's voice is full of contentment, and I want to cry because I could see Manuel looking upon his beloved island and hear the waves of the ocean calling him home. I've already cried a lot in this session. Instead, we schedule our next appointment, with Father Jose kissing me on both cheeks before parting. He asks if I have a way to get home, and I assure him I do.

There is a misconception that returnees long to return to live in the US, but I long for mobility—to ride in an airplane and see my family, to experience snow again, and then to fly back home to México and sleep in the sway of my hammock, my feet tanned with flip-flop marks. My life post-ban includes fantasies of getting a visa and spending Christmas with my family, of hugging my nieces and nephews so close I can see the dimples on their cheeks as they laugh. I picture myself washing dishes after our meal together as my family shares childhood stories, no video call time lag to work through. I dream of visiting friends and meeting their spouses, holding their babies, of joining together the parts of my American self and my Mexican self, no longer fragmented but whole.

A person bigger than a border. A home as big as the world.

Azul Uribe is a queer, bicultural writer, deportee, and speaker whose words have been featured in *Los Otros Dreamers*, *The Daily Dot*, the "Texas Standard," and other publications. Born in Cancún, México, she was deported in 2009 under the Obama Administration after fifteen years of living Stateside. She writes about post-deportee life, with a focus on reframing the immigrant experience to be a more inclusive and diverse space. She is working on an upcoming memoir and currently resides in Mérida with her cat, Chloe.

féi hernandez

POEMS

After Sappho

With chariot yoked to thy fleet-winged coursers

amá dragged the misfortune of love and Bellatrix with us,
our Pi'ma relatives didn't recognize us from their South
West across the Northern border—

we were wild Rough-Legged Buzzards desperate to
remember our migration trail.
Stratus clouds wrapped us with godly arms, our tiny visas
under our tongue, visas the size of el Sagrado Corazón
de Jesús prayer cards.

I met the wet disease Rheumatic Fever at 8. No health
insurance, but amá
took me to Harbor UCLA Medical Center, praying
there was enough sand to dry me.

I was too Abronia Angustifolia to survive
migration to Los Angeles.

Mami couldn't carry the dangling bones of me into the hospital,
but she never stopped apologizing to all the soaptree yucca she kicked to get

here.

Our chariot was not made of gold, only Sage Thrasher feathers and ocotillo flowers.

I know that Mami can be Cerro Mohinora in its fruiting days, angry and full of lava,
but as of late, she is extinct, her feet drag, no longer capable of running. She is

fume,
a volcano with no exit,

dormant life.

Migration that kept our last breath reserved for emergencies.

Migration that I only recall in cellular memory, but that lechuguilla tells me about in dreams.

A tear down my Mexican passport gives me permission to return.

What do we give up to return?
Alhena is so far, but she instigates my inner workings to return
to a time before body, a shaded purple place where my limbs were nebulas, my eyes,

starlight.

Cirrus clouds brushed the sky when I, the prodigal daughter,
was pulled out of wounded Earth—a simple slice across her stomach opened

me to Mami.

I didn't need Aquila's hoaxing to lure me back into my

yellow and incubated
jaundiced body.

I wanted to weep. I wanted to wake the world. I wanted to reconstruct my body
the way maps do. Line after line, a lie—Madera, Anahuac, Cuauhtémoc, a falsity—childbirth,

mine
to rearrange.

Did I fail to yield like Vega in a past life?

Nation-state governments my punishment?

I once flew over the Barrancas del Cobre as a Chihuahuan
Raven after amá and I argued one
dreadful waxing crescent moon after waning gibbous doom.
I always made it to the other side of the land's hurt,
thirsty.

Amá gave me the space to call back my wisps of hair across
the desert; she gave me all the
time.

By a river on the other side, a Tarahumara woman told me I
was mixed with ash. She said

the only way to understand justice—the devil and an
ancient God,

the same.

I don't remember my car seat, the petroleum stench across
the roads, border patrol. A third plane?

I remember growing up like Achilles, Nikes and corduroy
pants, fresh overalls and '90s patterned shirts.

America was always mine.

féi hernandez

How We / Tell Stories \

the mountain's rage / a curled child / rocks himself / to and from \ is / my sky / the horizon \ my nameless mucousy brother / coils / the land \ his unborn red heels / kick-buzz the land \ i give life to vibrant bones / fleshless wounded earth \ to my right / a rear view mirror / a cluster of gray hair / spider webs flap with desert wind \ past the passenger seat window / the pregnant boulder of a glistening mother \ the rock holds her milking nipples \ milk won't come \ the father cloud / a thunderstorm swirling towards the ant / of us \ how many times do we have to battle a dust whirlwind to make it back? / i don't dare roll my window down for fear the wind would sequester me / teach me how to fly \ i promise it's not an exaggeration / how many hours till we make it? \ our windshield chirps with the rain of sand \ the sky is a bruise that wants to take me / but don't believe anything i tell you / it's all a hot mirage hovering the black tar of the road \ my furrowed / focused brow \ the car's vents sigh an unnatural cold whisper \ my / native land / a furious songless abuela / that lost her voice after years of belting \ the evening strip / a dirty umbilical cord / tumbling across an endless blue \ our south star / pulls us home-home \ i should help amá drive / but she's already three sweaty cokes in / her blood / a boiling ponche / gurgling stories / a forced sticky joy / i should google the name of the cloud riding alongside us / a daytime ghost / i'm sleepy / mami is no longer driving / she is a million doves / jenny rivera is an octave higher than her hurt / i don't interrupt her ascension / her conjuring / she is booming voice and wings / grips the wheel as though it was a bottle / swerves / the poor roadkill \ don't trust me when i tell you there's a reason for the road trip \ translucent people walk the desert \ we held our breath crossing / over / or through to / bienvenidos a méxico / legally \ amá takes off a rough skin / the carcass of memory / her first swerving crossing \ over / or through to / a dream she walked in / until she awoke \ our silence dawned before us / there were no more songs to sing / sad album after album / of northern folk songs \ ave cautiva / or / por las calles de chihuahua \ a smell of damp desert wraps the wheels / a smell of gunpowder welcomes us to / our / ombligo \ a tired dog / our car / barks / our arrival \ abuelita / the sweetest atole / greets us from behind the porch's metal gate / she is a recurring waving gif / scared to be bit she stays afar / her mind hadn't moved since the last time we pulled / up / on her block / she forgets / the way amá stole me from her a life ago \ to love us \ i left her too / i think i had a choice / in a previous life / but don't believe me / abuelita and i can agree we're writing a new story \ inside the turquoise fossil of their home / abuelito slurps on the bone of an animal / under lamp light / furrowed brow / greets us / a rehearsed smile from memories in anáhuac \ everyone is diagnosed with the same disease / when we meet for the first time in months / everyone's chest glows to the rhythm of a soft pulse from the medicine of time / feed time / time / dale tiempo al tiempo \ i'm telling you \ don't believe me / i don't believe your grandma \ e(i)ther / that's what makes me love her

an old porch light / flames
the darkness of streets we hadn't been ready to walk \
the high
peaked silhouettes / the ghosts of mountains / still surround
our story / they whisper about
rain as tears \ a storm that passes / topples
mothers \ turns them into rivers / pregnant and all

i set my american passport down \ in chihuahua / the largest in méxico / i was hiccupped to life \ interrupted / you know what i mean? \ outside the window / a blue like drowned arms / legs / a foot here \ a head there \ the rose colored tail of sunset \ the father cloud returns / stomping the native ground / before the state massacred our people \ he slips behind the valley's shoulder \ apá stopped haunting us / when we became the hunters \ this is the beginning and the end \ this is how we tell stories / we start with the land you were born in \ trail the journey backwards / pace / fast / and / slow \ follow the people made out of stone \ hug the children that tumbled out of desertland fissures / listen to the mothers who lose / it / turn into rivers / fathers who shatter / fall / an eternal fall into abyss / like rain reversing back into the night \ or rest and remember \

Conception

After Janel Pineda's "In Another Life"

The migration never happened but somehow you and I
still exist.
Like desert rose,
we know only the memory of crystalline petals and not the
tumbling of betrayal that
created

us. Forget the almost gunned-down pregnant woman, the
jealous wife, the
caldo de res.
There is no bus to the maquiladoras.
There is no border. There is no drowning. There is

only ocotillo and tortillas hechas a mano. The pink house
and the backyard wall decorated with broken blue and
yellow tile.
A heaping plate of buñuelos and Abuelita's canela café
waiting to be
disappeared into our bellies.

In this life, our people are not things of Chihuahua City but
whole
worlds living amongst the Sierra Madre, Basaseachic, and
Samalayuca.

Everywhere, we are pronghorn antelope when we tire of
our anatomy—
playing unhunted, fully skinned animals, fed.

Tronadera does not mean gunshots and tunas bloom in
every
place water gallons are
left behind.

My name will still mean faith, this time in language
that recognizes me,

a language I know how to speak. My grandmother is still a
singer although I am
not a writer. In this life, I don't have to be. This poem
somehow
still exists. It is told

in my mother's native songs and she makes worry dissolve
like lime in
warm honey, throat
refreshed and free of silence. You and I do not leave each
other
across the border, grasping

angrily at each other's heart in search of *stay*. We meet in the
grassy fields
by Lago el Rincón, my arms overflowing

with mojarras del Siroco and Vicente Fernández tapes, and
together we graze the clouds
passing before your 1966 Ford pickup's windshield. We
watch sunset fall over

Naica we call our own and don't fear its stripping. I bite into
your lips, softly, sucking
color from substance, nibbling the memory of today's
conception.
Our laughter echoes

across the surface of the lake, one where we don't fear our
finding. We do
not have to hide here.
We do not have to hide anywhere.

---

féi hernandez (they/them) was born in Chihuahua, México, and raised in Inglewood, California. They are a trans nonbinary visual artist, writer, and healer. féi is the author of *Hood Criatura* (Sundress Publications, 2020). Their writing has been featured in *Poetry*; *Oxford Review of Books*; *Frontier*; NPR's "Code Switch"; *Immigrant Report*; *Nonbinary: Memoirs of Gender and Identity* (Columbia University Press, 2019); *Hayden's Ferry Review*; *The BreakBeat Poets, Volume 4: LatiNext*; and *PANK Magazine*; among others. féi is a spiritualist who utilizes a decolonial approach to ancestral energetic healing for (TGI) BIPOC. They collect Pokémon plushies. féi is the president of Gender Justice LA and is a cofounder of the ING Fellowship.

SURVIVAL

Girum Seid Mulat

POEMS

When I Dream of Mother(land)

i.
Inshallah
Sun rays push the town closer together
Elders bless us like every other morning
Right palms cup foreheads
Bekeñ Yaweleh
Let God keep you on the right side

My mother circles the table where
Harar coffee boils to blend
into the black clay jebena cradling
the steaming brew/perfumes
the air with notes of nuts & earth mixed
with fire to burn a melody
throughout Debrezeit/Dreams float
with the singing steam, becoming
real with every bite of bread, buna kurs, on the side

& I left
running
from jail cells that prayed
my name like men I did not know
asking my mother where I was hiding/
I left
running
from the burning embrace gifted
by a dictator's arms/Pushing
me away from my mother's home/
The one by the lake where the tenadam grows/
Where I sat
on a three legged stool/
I left
running
to become a tumbleweed crossing ocean by wind

ii.
It is December/Right foot stepping
off a plane & left foot searching
for Ethiopia/My face scratched/
Lung bruised by the icicle gust/
My mind still stretched
around my mother's round hips, letting
my cedar cheeks melt into her cinnamon skin/
Hoping to hear my mother's voice ask
Leje metah? My son are you here?

A bitterly cold basement my only shelter/
Three roommates/One bathroom/No bedrooms/
Months pass/
I wake to work 12 hours a day/
I wake to wish for a real job/
I wake to read messages from relatives asking for cash/
I wake to worry about my papers/
I wake to question the questions on the I-589 application/
I wake to think where I will find the money to file my case/
I wake to understand immigration is expensive/
I wake to decide I should skip lunch & send money back home/
I wake to wonder when I can go back home/
I wake to ask is this all worth it/

iii.

& only in sleep I can feel the walls dancing/shrinking
onto me as I listen to the elders blessing
my sunrises/I smell
roasting coffee mixed with my mother's fir fir
& fresh injera/But I blink too soon/
& the elders are not there/
& my mother is not there/
& now it feels like a dream/

12 hours working becomes 8/
Less than minimum wage becomes a salary/
Lunch or Western Union becomes both/
Buses become buying a car/

Renting a bedroom becomes owning a house/
I'm completely alone becomes I have a family here/

& still my case is pending/
& still my town gathers for coffee/
& still my mother makes fresh injera/
& still I miss my mother/
& still I'm dreaming/

What can be better than a motherland?
Sings me to sleep

> *Man ende enat, man ende hager*
> *Yet yegegnal mesoso ena mager*
> *Man ende enat, man ende hager*
> *Yet yegegnal mesoso ena mager*

Girum Seid Mulat

From an Ethiopian Child Living in America

An African country overflows with uranium & oil
as opal signals wealth to foreigners
There is a pool of gold the size of 10,000 football fields
It is lined with diamonds, but my father works
from dawn till the crickets chirp
& the frogs croak, alongside coworker amputees
On my way to school, I pass
clubfooted classmates cursed by the pesticides
a gift from Chinese & Hollander farm owners
who breed rainbow roses for our dictator & their lovers
across borders
I never saw the harvest firsthand
The gold, the diamonds, the oil, or the opal
I never saw the uranium, but I did read of the toxic clouds
it carved setting the sky ablaze
I did run from the gunfire
I did watch Muslims & Christians eat separately
I did see Gambella people pushed back
for their blackness like lepers isolated for their skin

& in America,
a white woman refuses to accept $10 cash cradled
in my right palm. A cop stops me on my street
What are you doing here?
Is this your driveway?
This is your home?

George Floyd's portrait & Hachalu Hundessa's songs burn
in my neck, pumping my heart faster. Our skins hexed.
My English doesn't bloom. It breaks
at pronunciations my mouth has not harvested
yet. Then, I saw cotton fields for the first time less than a
year ago.
Summer dying in South Carolina. The sun backing into
concrete clouds as snowballs crowned stems the color of
coffee.
They alone & only the breeze there to pick at them.

Now, a corner office is labeled
with my name. Colleagues doubt that I am their peer.
Go back to your country replaces *Have a nice day.*
It has been six years, but still I do not know freedom.
& I realize that being both Black & immigrant
is a body dipped in cinnamon & hardened with cedar.

Girum Seid Mulat is originally from Ethiopia and has established himself as a dominant figure in the arts scene of the Amharic-speaking community. His debut poetry collection, *Passengers of the Coin* (2019), was released in both the United States and Ethiopia. His work is a testament to his creativity and attention to detail, while also playing as an ode to tradition and a nod to the nostalgia East Africans feel. Girum currently performs in the greater Washington, DC, area, participates in speaking engagements throughout the community, and serves as an activist for immigration rights. Girum is also the founder and president of the Ethiopian Arts Society in North America.

Aline Mello

Fit

1

I am taking up space that belongs to an American citizen.

I don't remember the first time I heard this, but it has become a burden I carry. In college: I am taking a spot that could've been a citizen's. At my job: taking up a position that should be filled by a citizen. But it is my weight that offends, that makes it so much worse. If I'm going to be in the US undocumented, shouldn't I at least be smaller? I should take up as little space as possible—maybe if the American citizen and I are each thin enough, we can both fit.

My weight trauma was born in Brazil and raised in the US, just like I was. Back there, when I was seven years old, Mamãe would watch my food intake, and the aunts and church ladies often talked about how I'd better lose weight before I became a young lady. My mother's obsession with my body was a reflection of her own trauma. As a young girl, after her mother passed, she went to live with an aunt who wouldn't feed her if she felt my mom looked chubby. Food wasn't sustenance but reward. With her daughters, my mother's neurosis found new targets. My sister and I were drinking diet teas and taking diet pills before we even hit puberty.

When I was in sixth grade, now living in Georgia, I stood in line with my class to check out some library books. Someone smacked my behind, and I turned to see Bruno 4, the fourth and worst Bruno to join our grade, looking smug and almost angry.

"Por que?" was all I could say. The only people who ever smacked my butt were my parents and sometimes my sister when we were playing.

Bruno 4 shrugged. "You have a big butt."

"Oh," I turned my head to look at it as if noticing it for the first time like an extra limb that had grown overnight.

From then on, I couldn't stop thinking about my butt. I started tying sweaters around my waist. I compared it to my friends'. The words "fat ass" haunted me. In Brazilian culture, having a big behind is normal, but I was no longer in Brazil. I was in the United States, where a bigger butt meant you were a fat person. Nobody wants that, no matter their culture. And being undocumented while in a larger body underlined the issues.

Restraint. Invisibility. These were things to work toward both in my body and in my existence because of my immigration status.

To fit in means to be small enough.

2

My mom signed me up for Weight Watchers when I was in eighth grade. I was the youngest member in the group meetings, and I was embarrassed to tell my friends at school about it. There were moments when I didn't think there was anything wrong with my body at all, that maybe my size was normal. After all, I rarely ate junk food, didn't drink soda, and never ate out. But those thoughts were overshadowed by calorie counting and the search for some ideal I couldn't really visualize.

In high school, Mamãe decided she would no longer buy me my correct size of clothes. Instead, we would buy pants too small for me, an incentive to lose weight. I explained this reward system to my friend Jessica once, and she looked at me like I was crazy—not innovative and smart. So, I stopped telling people about it. Losing weight was good but talking about losing weight was not.

If I wasn't participating in every new fad diet, I was aware of them. With the encouragement of my mother, I read the books, took the diet pills, measured my rice and beans, made sure the piece of meat I ate was no bigger than the palm of my hand.

Of course, the weight-loss tactic proved futile again and again. I'd lose the weight, but I'd gain it all back and then some.

My closet was full of too small clothes that taunted me.

3

Growing up undocumented in Georgia made me crave security above all else. The state's strict anti-immigration laws are there to make life so intolerable for immigrants like me that we self-deport. For example, when I was a teenager, the Board of Regents decided to make every undocumented student pay out-of-state tuition for higher education. Even when President Obama signed DACA, Georgia leadership refused to change its policy. Fortunately, I was able to afford some classes at a community college after high school. When I went for orientation, the advisor tried to persuade me to take ESL classes, even though I had AP English credit. They ended up making me take the test to prove my English proficiency, though they knew I'd been in the US public school system since second grade. At every turn, higher education felt like an impossible feat.

In his campaign ads, our current governor gets into a pickup truck and says, "I got a big truck, just in case I need to round up criminal illegals and take 'em home myself." He turns on the ignition, smirks into the camera. "Yep, I just said that."

I have never felt I belonged in the halls of higher education—not even at a community college. This country and this state have made that clear to me. Any time I have been inside a room of higher education, a conference, a space where my parents have never been, it was because I squeezed my way in, brought my own chair. It's hard to imagine the system changing. Even though our US senators now are both Democrats, Georgia's local government is still Republican and anti-immigrant. It's always been easier to imagine I will change instead. I can practice speaking without an accent. I can keep my head down. I can become thinner, smaller, and less threatening.

I can make sure that they have no reason to be afraid of me.

4

My eating disorder became more pronounced in college. Mamãe started paying Kelly, an unqualified White woman from church who called herself a trainer, to keep tabs on my food and exercise. She would mix Bible verses with her training, reminding me that thinness was a sign of self-control and being fat was gluttony. I repeated 1 Corinthians 9:27 like a mantra: "I discipline my body like an athlete, training it to do what it should. Otherwise, I fear that after preaching to others I myself might be disqualified." A fat person cannot talk to others about God, the trainer said. Their fatness is a visual marker of sin. Being thin was holy and being fat was not. With her guidance, I became my smallest—a loose size 9.

I weighed myself every Saturday and sent Kelly the results. If the number didn't move down enough, I would sob, my shoulders shaking as if I'd lost a loved one. Saturday mornings would determine whether I had a fun weekend or locked myself in my room away from all food and people.

One successful Saturday, when Mamãe was sitting at the computer in the hallway, I said, "Look," showing her how my pants were finally fitting me. "The plan is to get to size 6 so that if I gain some back, I'll still not get to double digits."

"Minha flor," she said, looking at me. She reached out and touched my waist. "If you decided to stop losing weight right now and just stayed this size, you'd be perfect."

I couldn't breathe. I rushed to my room in confusion. This—my mother telling me I didn't need to lose any more weight—was what I'd been waiting to hear my whole life. Right? Why didn't I feel any relief?

I looked at myself in the mirror. She'd said this was good enough. It was perfect. My thighs still touched, would always touch. My pelvic bone stuck out just barely. When I lay on my belly to sleep, I felt my ribs on the mattress. My butt was still big, but no one would call me a fat ass. I wanted to run outside and scream or hit something with my whole body. I wanted to release the rage that had built up inside me. I swallowed it instead.

The weight came back, of course. All of it and more.

5

My eating disorder was never about vomiting. Oh, I tried. But I was a singer and read about how throwing up damaged the vocal cords, and as a good Christian girl, I wasn't about to knowingly damage my instrument. I don't vomit up my meals, so it's difficult to tell when the symptoms of the disorder are returning in recovery. They are as learned as reflexes. And they feel good.

Sometimes I feel hunger pangs and joyously ignore them. I wait until I feel hollow and light. When I drink water, I can feel it sloshing in the emptiness. It must be how a bird's hollow bones feel—the airiness, the pain, as if being stretched upward toward flight. When I wait that long, eating feels uncomfortable. Anything solid feels like stones weighing me down.

On Brazilian thinspiration social media accounts, there's a fad called barriga negativa—*negative belly*. The women—and they're *always* women—take pictures with a sports bra, showing how their ribs jut out more than their stomachs do. The skin dips, then comes back up over the pelvic bone.

Like the curve of the moon at its thinnest.

6

I am taking up space that belongs to an American citizen.

The first time I found out I had an eating disorder, I was in a therapy group and the therapist said, "Everyone in this group has an eating disorder" in a matter-of-fact tone. Everyone around me nodded like they all knew. But I felt exposed. An eating disorder sounded like a vain, laughable phase a teenager might go through. Eating disorders leave women emaciated, bone draped with skin. But I wasn't even normal thin. I hadn't even been good enough at having an eating disorder.

Surprisingly enough, if Donald Trump had not become president, I might not have started my eating disorder recovery. In 2015, when he started his campaign, I decided to be open about my immigration status. I started telling close friends, then acquaintances, then posted about it on social media. I believed that people just didn't know better, and the more educated they became, the less hateful they would be. I especially thought this of my Christian friends and coworkers. Surely, a faith that had brought me up to be hospitable and to extend kindness to strangers had taught those same values to my fellow believers? However, the election results proved me wrong. It sent me down a spiral and made me seek out professional therapy, not just Christian counseling.

In therapy, I learned that eating disorders don't always mean weight loss or weight gain. They don't have anything to do with food, not really. In my case, my disorder came from the need to control *some*thing. To have a say in some part of my life. And the desire to be thin—yes, even emaciated—was a need for invisibility. I wanted to be able to fit under my bed. I wanted to hide behind a pole, sneak past in a blur like stagehands do when changing the scene in a play. This was safety. To be undetectable is to be a safe immigrant in the United States.

But my body has never allowed it.

7

I belong to whatever space I am in simply because I am in it.

This is still something I'm trying to learn. Growing up undocumented, from age seven to the present, caused more harm than I could've anticipated. I'm still going to therapy regularly, and my therapist speaks of my being undocumented as complex trauma—an ongoing, wide-ranging abusive traumatic experience that causes long-term effects. Having it classified as such validates the real issues it's brought on. After all, my eating disorder cannot be separated from my being undocumented.

After years of therapy, I understand now that there is no perfect version of me that this country will accept. No number on the scale. I will never be small enough. But perhaps my largeness is supposed to make me less apologetic about taking up this illicit space. Like my body is rebelling before I do. Maybe it's a "fuck you" to the Man. Giving in, I could spread myself larger, dress in only big prints, horizontal stripes, flowing fabric, nothing holding me back. I could reach with all my sides to the world around me, no belts, no walls, no borders, breaking seams and rivers.

Maybe this is how I'll fit.

Aline Mello is a writer and editor living in Atlanta. Her debut book of poetry, *More Salt than Diamond*, was published in spring 2022. A Brazilian immigrant, she is an Undocupoet Fellow, and her poetry has been published in *The Georgia Review*, *The Indiana Review*, and *The New Republic*, among other publications. She devotes much of her time to serving immigrant communities and preparing undocumented students for higher education.

t. jahan

Any Day Now

Rumbling like an approaching train. I hear the heavy
knocking on the door. They found us. I climb out of
my window in a tattered nightgown with dried blood spilt
on purple roses. Squeeze my small doughy body through
the gap above the clunky air conditioner.
Traversing the rusty fire escape, escape, escape, I look
over my left shoulder. They're still after me. Skipping
balconies, and trespassing homes, rooftop to rooftop, I
look over my right shoulder. Where is my family?

I wake in gasps.
The same dream again, and again.
I am four years old,
maybe five,
maybe six,
maybe twenty.

Any day now, any day now, I'll be going
back to where I came from.

[<<] FLASH [<<]
Passport-sized photographs
Don't smile. Don't dare blink.

[<<] BACK [<<]
to when I came
Sign your name like this. Stand straight like this.

Mary Janes on my feet, reversed.
I am eight.

For the first time, my shoes are on the wrong feet and, sigh, I am standing for hours. But shush, not until the immigration powers-that-be give me permission to sit, rest, breathe, and speak.

Not yet, not yet. Neither witness, nor victim.

But PRESENT
(I shout internally to my third-grade teacher who has finally learned how to pronounce my name)
though I am marked ABSENT from school.

Mami and Papa, cast their long shadows behind me
as I stand before the immigration judge.

Are you a boy or a girl?
I hear in annoyed, inconvenienced English.

তাড়াতাড়ি, quickly, put a headband on that child.
I hear in anxious, fretful Bangla.

Because they get to decide, they get to decide,
they get to . . .

[>>] FLASH [>>]
[>>] FORWARD [>>]
to that night,
the next night, and every night after that.

I lie in bed, eyes winced shut with earnest, clasped hands. I am tired, yet need to make my case to The One. Hear me, I implore You. Please, please, please, let me wake up somewhere else. Anywhere else. Animate and transform me into the main character, a young girl in Kanto named after cherry blossoms. A girl with warm, ambitious heart, who will claim her wings and the right to her destiny. Lover and beloved of the people. Protected yet free to fly.

The same dream again, and again.
I am nine years old,
maybe ten,
maybe eleven,
maybe nineteen.

Because imagination is no longer doing it for me. Because the library lets me take out only 25 books at a time. Because storybooks don't have ever-afters. Because it's Sunday and public television is trying to sell me wondermops and superfryers. Because the flickering red light on my gameboy tells me that the batteries are about to die.

So, let me go to sleep and wake up somewhere entirely different, and let it be permanent.

[<<] OVER [<<]
[<<] & OVER [<<]
I am three and a half.

Somewhere in the corner of a photograph, I stand in an overcrowded New York City apartment, unsure of where I belong, having been transported from a cold, beautiful country in Europe that has given me no birthright. And in the center sits sweaty, smiley Papa, the office salaryman turned odd-job laborer. Cradled by his firm forearms, an anchor baby and the holy grail passport.

[>>] SKIP [>>]
to when I am seven.

Mami and Papa teach me and my younger sibling with a sense of urgency:

1. *Friends are not important.*
2. *Don't get too attached, we may have to leave.*
3. *We must send all the money beyond this place that may not keep you.*

[>>] SKIP [>>]
to when I am ten.

Forget about feel-good things like nintendo 64s, piano lessons, and poptarts—we need to drain water out of the motherland and build an edifice somewhere in a crowded city called Dhaka and call it The Tower of Dreams.

[ ] SKIP [ ]
to never.

Never will I live in the Tower though my pictures will adorn its walls like royalty. The noxious dreams fueled by entrepreneurial ’murrican spirit that built an escape, a hideaway, in the guise of success. How romantic, how it oversees a moat of tin slums.

[>>] AGAIN [>>]
[>>] & AGAIN [>>]
I am seventeen.

On a steep concrete hill, moments before a hopeful departure for the motherland, begins a hostile contest for power with the beloved man I once called Papa. I am wrenching at the holygrail,

let it go, let it go, let it go. And he is tugging at the property rights to the Tower of Dreams,

give it back, give it back, give it back.

Papa fled before the police came. *You stay here, and I go there*, a violent tradeoff with no winners. An empathetic woman coaxes out the misdeeds of our dysfunctional family. I am sent to the familiar emergency room to mend my crooked pinky. Another Order of Protection leaves Queens Family Court to reach someone with relinquished citizenship dreams.

[>.<] HOW! [>.<]
[>.<] & WHY? [>.<]
I am eighteen.

Mary Jane in my head, reversed.

Cravings and questions envelope me. Trickster spirits find fertile ground in my brain and take root. Dreams, ungraspable, going up in smoke and vapor. I am in a chokehold. I reach out for help, should the therapeutic hands slip away.

[>>] & THEN [>>]
[>>] & THEN [>>]
I am twenty-one.

And I know I am one of the lucky ones, still.

Somewhere in Long Island, I become a naturalized citizen.

Somewhere in the Tower of Dreams, I become an older sister to a half-brother I have never met.

[__] & NOW [__]
[__] & NOW [__]
I pause my mind and slow
[>>] [.5x] [>>] my breath.

They say our souls have been charged with all that they can bear. What if I picked this life like a kid in a toy store with no budget: *That one. That one over there. There's going to be a lot of shit going down. Can definitely work with this.* What if this was the world I received entry into after an earnest request?

Indeed? One may ask in bewilderment.

Indeed, I ask for you to consider.

Because I prayed when I didn't even understand The One, and asked to be a fucking cartoon when I was nine. Because I asked to defy nature and physics when I was ten. Because what I sought for was escape, for respite when I was eighteen.

But that would have lifted me from my journey, my purpose, my life.

Press [>>] fast forward [>>] on my soul,
I have woken up, hands still clasped, animated
and transformed.
It sucked sometimes, I know it wasn't all my fault,
but what if . . .

I climb out with

purple roses. my small

body above the air

Traversing the fire

Skipping

and passing

I wake

I am four years old,

maybe five,

maybe six,

maybe twenty.

now, now, *back to where I*

came from.

[<<] FLASH [<<]

smile. dare blink.

Sign your name Stand

straight

on my feet, versed.

For the first time I

sit, rest, breathe, and speak.

I stand before the judge.

Because they get to decide, they get to decide, they get to . . .

---

t. jahan, a part of the desi diaspora, is a writer and teacher dreaming of transcending politics and language barriers. These days, they are writing a novel set in 1970s Bangladesh, a children's book, and narratives and memoirs to bring awareness to Asian immigration and diasporic experiences. Their work has been published in *The Margins*, PEN America's anthology *DREAMing Out Loud*, the New York Writers Coalition's literary webpage *The Journal*, and more. They are self-teaching themselves piano and take aphorisms on satchels of tea seriously.

Julissa Arce

Not of Their World

My fellow Goldman Sachs analysts and senior colleagues were flabbergasted when I told them I had never been skiing.

"WHAT! Are you serious?"

"How is that possible, J?"

They couldn't imagine that someone hadn't grown up spending a small fortune vacationing in Aspen. Many of them still had their parents' credit cards or were living rent-free in spare Manhattan apartments, whereas I lived in a one-bedroom apartment with a roommate. I wasn't about to tell them that as a child I was lucky to go to the dirty beaches in Corpus Christi, Texas, on the Gulf of México, and stay at a Motel 6. When my family could splurge, staying at La Quinta Inn was, like, *wow!*

Wall Street was a world of fancy parties, the trading floor, business trips that included stays at five-star hotels—so different from the one I grew up in—but I knew the importance of fitting in. Success in America didn't come from hard work alone. I had to learn how to blend in. Even though I had the same job as my white colleagues, I was often mistaken for the office assistant. I shopped at the same high-end stores, but I was met with suspicion. Once, a racist store clerk took a necklace out of my hands and asked, "Do you know what Prada is?" At a team outing at a fancy restaurant, I was asked by a group of white people to bring them water on my way back to my table after using the ladies' room because they thought I was the waitress.

I didn't want to get caught up in the shallow Wall Street rut of money and more money, but I had to figure out how to act rich and white. I was tired of constantly feeling awkward and out of place.

But learning to fit in to this world was harder than I'd thought. In the summer, my boss took a few of us out to the Hudson, in the New York Harbor, for a sunset sail. Having never been on a sailboat, I got seasick and spent most of the afternoon trying to keep my food down and drinking ginger ale. Maybe it was a good thing I spent most of the time in the hot bathroom since that gave me an excuse to skip out on the conversations about spring sails to the Bahamas or complaints that sailing wasn't an official college sport.

I needed a way to connect with my peers and blend in during social situations. I thought about all the activities I had heard my colleagues mention, or ones I had seen wealthy white people do in movies. I thought I might like tennis, but lessons were too expensive. I had a bad back, so riding horses, another out-of-my-budget experience, was also a nonstarter. When my colleagues told me of their winter holiday plans, the hot tubs at the end of a long day of skiing, the wine by the fire, the exhilaration of gliding down the mountains, I decided that skiing was the thing to try.

I knew my friend Harry was the right person to ask to join me on my first skiing adventure. He was an avid skier, I was able to be myself around him, and I trusted him.

"You're going to love it!" he assured me.

As I struggled to find my footing in such a new and different setting, Harry was a godsend. We met in 2004 during our summer internship, and now, a year later, we were both analysts at Goldman Sachs. At a networking function, when someone complained about the wine, he whispered to me, "It's free wine, who cares?" And I noticed him taking his used glass to a tray instead of leaving it on the table for the waitstaff to clean up.

He was from their world, but not of their world. He had worked construction every summer in high school to save for college—not because he needed to, but because he wanted to. When I told him I'd sold funnel cakes to attend the University of Texas at Austin, he was genuinely impressed. "I just didn't want to ask my parents for spending money," he said. "But you paid for *all* of college!" When he shared that his grandmother lived with his parents, I told him, "That's pretty Mexican of y'all. We don't put our parents in homes."

Harry took me to horse races at Belmont Stakes where I wore a big hat and a pastel Sunday dress. I felt so fancy sitting in the stands watching as horses went all out, taking long strides down the track. We made some bets, and beginner's luck struck me twice.

He taught me the way of America's elite in small ways. He insisted we watch hours of classic movies and TV shows, joking, "How could you say you've lived otherwise?" I had to admit, sitting in the living room of his tiny Manhattan apartment watching *Casablanca, Butch Cassidy and the Sundance Kid, When Harry Met Sally,* and every single episode of *Seinfeld* did make me feel newly born in America. References I never caught before suddenly made sense. An ex-boyfriend once said to me, "We'll always have Paris," and at the time I'd thought, *But I've never even been to Paris!*

Harry was helping me survive in a place that constantly left me feeling like an outsider no matter what I did. Watching classic American movies, developing a taste for expensive wine, wearing clothes I could barely afford, I thought were necessary things to help me fit in, to find unquestionable belonging.

When I arrived in the US at the age of eleven, I thought learning English would help me belong in my new school. I had practiced enunciating my words to match those of the blond girls in my classroom. I imitated them so well that people later told me I sounded like a white girl, something I took as a compliment. But then one day in eighth grade, we were reading about animals and the word "crocodile" appeared on the page. I pronounced it "cocodrile," resembling the Spanish word "cocodrilo." My classroom filled with laughter. It didn't matter that I was an honor roll student: one bad move and I was back to being the dumb Mexican girl who couldn't speak English. I still remember how the humiliation burned my face. I didn't want that to happen to me here at Goldman Sachs.

One Saturday morning, before the sun poked its head up over the high-rises, Harry and I met at Union Square to hop on a bus to Hunter Mountain in the Catskills of upstate New York. Harry showed up with a bagel, a pair of extra gloves for me, and some handwarmers I didn't know I'd need.

Once on the mountain, tiny white children in their adorable pink-and-white onesies swooshed past me. Their pale faces and red checks glowed with excitement. Meanwhile,

I had squeezed into my friend's small ski clothes and looked like a frozen stuffed sausage in her blue pants and white jacket. I'd saved for weeks to be able to take the trip with Harry but didn't have enough to buy the clothes to fully play the part. My annual salary as a first-year analyst was $55,000. That was a lot of money for a recent college graduate, but not as much as I needed to live in New York City and to take care of my parents who were back in México. At work I was eating at Michelin-rated restaurants, boarding yachts, and shaking hands with millionaires, while at home I was scraping to get by. My roommate and I oftentimes shared a Sbarro's pizza special of two cheese slices and a Coke for dinner to make ends meet. I wondered whether these children on the mountain knew how lucky they were to grow up being part of this world—not having to be self-conscious, not having to change themselves to fit in, and not having to feel like a sellout for wanting to fit in.

The instructor, who looked like he belonged in an episode of *Saved by the Bell*, kept calling me Julie despite my best efforts to remind him my name is Julissa. "It's like Julie and Lisa combined," I insisted.

"Okay, Julie," he'd respond.

It made me wonder whether people at work called me "J" because they couldn't remember how to say my name. Who knew? In that moment, I needed to stay focused on not dying in my early twenties in some freak skiing accident.

"If you get scared, just bring the front of your skis toward each other. Make a pizza wedge! That will slow you down," the instructor said before releasing us for our first trial run.

I was glad Harry was not there to witness my embarrassment

as I ate snow for breakfast. I had persuaded him to go off and do some "black diamonds" (a term he taught me on the bus ride to Hunter Mountain) while I learned how to properly put my boots on.

After the baby-faced instructor was done walking me through how to strap into my bindings, slow down, and get up when I fell, I was officially ready to hit the slopes. Harry was waiting for me at the end of my lesson. His nose and cheeks were as red as the children's who'd swooshed past me that morning.

"Are you ready? We'll get some good runs in before the afternoon sun melts the snow and it ices over," Harry said as we sat on the ski lift that took us to the easiest part of the mountain where we could find slopes as flat as 25 degrees.

"Uh huh," I said nervously as I looked down at the snow-covered pine trees, and at the dozens of bodies who swayed right and left as they elegantly made their way to the bottom.

Harry explained that one of the most difficult things was figuring out how to get off the chairlift. I did as he said, putting my poles in one hand and using the other to push myself off the chair. Then, as though I had done it a million times before, I used my skis as skates to navigate the flat ground toward the start of the run.

"You are a natural, that was the hardest part!" Harry said. I hesitated a bit before pointing my skis down the mountain. "Just look at me, and your skis will follow," he said. He skied backward as I kept my gaze on him and somehow my body moved in his direction. His helmet covered his brown hair, and his goggles hid his greenish eyes, but a smile warmed his face as we slid through the snow falling from the trees.

As I made my way down, I thought about the rocky state of Guerrero, México, where I grew up. My hometown of Taxco was built on a mountain and is surrounded by luscious green hills on every side. It hit me that I had never seen a mountain covered in snow, and here were dozens of them all around me. Some of the peaks were surrounded by clouds, and the sun made the snow look like glitter.

A rush of adrenaline, and a tender feeling of flurries, like I couldn't believe I was there, filled my stomach as I moved faster and faster down the mountain. I wanted to scream in excitement, but I reined in the stampede of feelings. My body demanded my full attention as it shifted right and left.

It wasn't until we were at the bottom that my heart exploded with triumph. I had just completed my first ski run! I raised my poles to the skies and let out a loud laugh that made the people around us look in our direction.

Back at the office, my boss had suggested I laugh less loudly. I found it impossible that over the noise of shouted profanities, dozens of phones ringing, and men pounding on their desks, it was the level of my laugh that was disruptive. And here on the slopes, the raised eyebrows and side looks told me my joy was offensive to those around me. I didn't care. They should have all been laughing and jumping for joy at the glorious experience of gliding down a mountain.

After a few more runs we went to a different chairlift that transported us to a whole new world of possibilities, the blue slopes. The higher we went, the more beautiful the views. The population of the slopes looked like the trading floor at Goldman Sachs: mostly male, white, and stiff. At this elevation, I wondered whether anyone thought I didn't belong, with my sausage outfit and loud laugh. Or did they assume, like my colleagues, that if I was this high up, walking among them, I must be part of their universe? I didn't row crew or play field hockey or squash. My parents hadn't attended college. They had been traveling merchants selling anything from jewelry to funnel cakes to support our family. Harry's uncle had been a partner at Goldman Sachs. Another of our fellow interns' dad was my boss's boss's boss. To be the first in my family to go to college in the US, to have this job and these experiences—it was what I had been working for all my life. And despite all the challenges of being undocumented, I was here doing the same job as my privileged colleagues. Now I had reached the peak of the same mountain, too.

But even as I marveled at the beauty of the whiteness around me, I knew I didn't want to get sucked under it. I didn't want my laugh to fade. I had already given up too much to have this job, this life that promised me financial security—but at what cost? I had already paid a high price to be in this country, growing more afraid each day that someone might find out I was undocumented and working on Wall Street with fake papers. I constantly felt like an outsider in this place that was my home. And what I didn't know that day on the mountain was just how much more I'd have to sacrifice. Like when my dad passed away in México a few years later. Since I was still undocumented, if I had gotten on a plane to be with him, to hold his hand as he took his last breath, I might not have been able to come back to the US. I might have died trying to cross the desert like so many others. So I stayed in my high-rise apartment in the financial district, chasing the acceptance of the rich white people around me. There isn't a day I don't wish I had taken a flight to be with my father. Job, money, and belonging in America be damned.

Going down the mountain, I felt an immense sense of sadness and happiness at the same time. I felt on top of the world—and how much more beautiful the world was from up here! As Harry and I did our last run, down the steepest slope of the day, I focused on the *swoosh, swoosh* of my skis as they cut through the snow. I was hooked. The rush of adrenaline pumping through my body was exhilarating, scary, and freeing.

The Monday after Harry and I went skiing, I was eager to share my experience with my colleagues. After getting coffee and peanut butter toast, I asked one of the other analysts, a woman with perfectly straight white teeth, how her weekend had been.

"It was good, nothing too crazy. I had dinner at Nobu Saturday night," she said. "How about yours, J?"

"I took a day trip to Hunter Mountain. It was amazing!" I said excitedly.

"You didn't think Hunter sucked?" she said, making a face. "The only real skiing happens out West, or in Canada. But for a quick trip, Hunter does the job. I guess."

As time passed, as I grew closer to my roots, I realized that no matter how hard I worked, how much success I achieved, or how many dollars I amassed in my bank account, belonging in a white world was not a dividend that came along with it. Each time I thought I had done what I needed to become one of them, I was reminded that dressing the part wasn't enough. And thank God for that because if I'd kept wanting to become one of them, I wouldn't have broken free. Instead, I sought belonging closer to my roots. I moved to El Barrio where eating tacos wasn't a cultural field trip. As soon as I stepped out of the subway and onto 116th Street, I found myself home. I read books written by brilliant Latina authors whose works were not taught in school. Eventually, I even left Wall Street. I did return to the mountains again, this time with my Latino friends. We were loud and it was glorious.

Julissa Arce is a Mexican and American writer and activist. She is the nationally bestselling author of *My (Underground) American Dream* and *Someone Like Me*. Her writing has appeared in the *New York Times*, *TIME* magazine, *Vogue*, and other outlets. Her most recent book, *You Sound like a White Girl: The Case for Rejecting Assimilation* (Flatiron, 2022), is a manifesto against the ways America demands assimilation yet denies belonging, and a rallying cry to create space to truly be ourselves. Julissa immigrated to the US from México at the age of eleven and was undocumented for almost fifteen years. She became an American citizen in August 2014. Before becoming an author, she built a successful career on Wall Street. She was named one of the 25 Most Powerful Women of 2017 by *People en Español* and 2018's Woman of the Year by the City of Los Angeles. Julissa lives in Los Angeles with her husband and their two cats.

Mariella Mendoza

Montaña a Montaña

In August 2014, I pitched my tent on the East Tavaputs Plateau. I was twenty-four. It was my first time camping. I've always been curious about climate justice and had signed up for this trip organized by Utah Tar Sands Resistance to learn more about US Oil Sands's strip-mining project. Southeast of Salt Lake City, nestled between the Uintas and Moab, the Tavaputs had been chosen as the potential drilling site for tar sands, a substance made up of more carbon than most other fossil fuels produced in North America.

The plateau, in danger of being destroyed, was beautiful; the tall, thin aspens made me feel as if I were inside a mushroom, their white trunks drastically contrasting with the green of the leaves, the open blue skies. A small creek divided one hill covered in dandelions from a second hill carpeted in aspen leaves. Unlike the Andes, whose mountains I had grown up walking, the Tavaput Plateau is dry, but being in this high altitude reminded me of those mighty rainy mountains.

Local activists from Salt Lake City had lent me everything I might need: hiking shoes, a sleeping bag, a small but roomy tent. I was used to sleeping on the ground but not with this much gear. It was a luxurious and unfamiliar feeling. That night we built a fire and swapped stories. They talked about tree-sits and protests; I told them about my life—my forced separation from my mother when I was twelve years old because of an abusive relative, my migration from Peru, and my expired visa. About leaving that abusive household only to become homeless at age nineteen, obtaining DACA status and finding stability for the first time in my life. The activists talked about their families too and asked me how I had become interested in social justice. I told them about my mom searching for me for years, unable to call the authorities because of my undocumented status and her own ignorance of US laws, the longing we had for each other, and the impossible distance between us. The activists and I stayed up talking until the moon rose. I had forgotten what it felt like to see stars at night, to be surrounded by infinite sources of light. It was dark, and the occasional breeze disrupted the warm night.

I made an altar with huayruro seeds and rose quartz. My grandmother used to say, "Nosotros nos acomodamos"—we accommodate ourselves, we find a place to fit, we build a home in spite of everything. I arranged my blankets around the thick sleeping bag as I read the instructions attached to the zipper; the small orange tent already felt like home. For undocumented immigrants, home is a strange concept. For a formerly homeless nonbinary migrant, home is a fairytale I didn't quite believe in at first. But that day, home found me—there, surrounded by aspens underneath a starlit sky. Facing a billion-dollar oil project that I would spend the next two years (and longer) fighting, I found home in defiance, in doing the impossible, in being held by the palms of the earth.

The next day we headed up the Book Cliffs past the creek, toward the mining site my new friends were here to scout. "You don't have to come along if you don't want to," they told me. "There've been people arrested here lately, so it might be risky." *It's riskier when you don't have papers*, I thought. I kept walking. I wanted to see for myself why we had come here. I knew what it was as soon as I saw it. A scar on the earth, deep as a creek, filled with a sinister black substance, shiny and tarry. The word *evil* came to mind. A hole in the earth that shouldn't be there. My eyes filled with tears as the activists around me gathered samples and took pictures. I could see in their faces the same disgust at being confronted with this monstrous crevice on the plateau. "Money," one activist said. "Money and oil go hand in hand."

We headed back to camp and got ready to return to the city. I picked up my blankets and held them close to my face; the smell of tobacco and campfire held on as I folded them softly. I picked up the sleeping bag and the makeshift altar I had made and said goodbye to the aspens. "I'll be back," I whispered. "I promise."

I spent the next two years of my life helping the local Tar Sands Resistance crew, camping onsite and facilitating workshops in the city. Then, in the summer of 2016, I, and most Indigenous land defenders whom I know, dropped everything and went to North Dakota. In late August our friends asked us to go to Oceti Sakowin Camp to drop off firewood and food, but within days of our arrival we realized there was no leaving. Since the beginning of April, local and neighboring tribes had begun gathering at the Standing Rock Reservation in opposition to the Dakota Access Pipeline, which was set to be built underneath the Missouri River and through Lakota burial ground. By September, the #NoDAPL movement's call to action had mobilized thousands to camp out directly on the path of the pipeline, including myself. My mom called me often and pleaded for me to go back to Utah, away from the police violence and brutality that targeted the water protectors, but I knew I was not going to leave.

I knew I needed to be there.

People from all walks of life in North Dakota were supporting the water protectors. Scientists swabbing tents, organizers assessing the damage caused by the chemicals police had sprayed on us, artists patching banners and doing onsite screen printing to help others spread the word. Because of my lack of legal status, I was often assigned tasks within camp and to help with media strategy. I've learned to pitch stories to media outlets with intentionality because I know too well how narratives can be misconstrued. Too often mainstream media outlets will focus on a singular person, or moment, instead of the relationships between our communities. By separating issues and ignoring how they are interconnected, they compartmentalize our struggles and our stories. I am not just a migrant organizer who focuses just on migrant issues; I can be a migrant organizer who also focuses on climate change and Indigenous sovereignty.

When people ask me why I joined this environmental fight and point out that this land is not even mine, they forget that fighting to defend land and to protect the sacred is not a singular fight, but one that is linked to many. By choosing to resist this pipeline, we are upholding our right to defend sacred Indigenous land, while engaging in practical forms of solidarity across tribal communities. If we look at land defense through an international eye instead of one centered on the United States, we can see that this fight is as old as colonization itself and that the defense of our culture is, too. All of our rivers are connected, including the Huallaga, the river that my grandfather grew up beside in the Amazon rainforest, and all of our struggles are connected, too often by the same people who are building pipelines, who conveniently profit from the oppression of marginalized communities. I cannot choose just one of my identities, and we cannot ignore the connections between our oppressors and the legacies of genocide. And if we are to live, capitalism must die. We cannot ignore the connections between climate change, environmental racism, large corporations, and the displacement of migrants across the world from their land.

My mother failed to see it this way, even though the same fight was being fought in Peru as the government seized the land of native Andeans to sell their water to other countries. She pleaded with me to return to my home. I always replied, "Home where?" How could I leave? In land-defense immigrant circles we have a saying: La tierra no se vende, se ama y se defiende—You don't sell the land, you love it and you protect it.

In just a few months Oceti Sakowin grew from a small campsite to a full-scale occupation, with flags from more than a hundred Indigenous communities lining the main road and several thousand campers who had all learned to cope in their own way with camping in that cold North Dakota winter. With temperatures below freezing, most of the tasks included chopping wood, lining flimsy tents with wool blankets to keep campers warm, and shoveling snow. Some days were harder than others. Near the end of winter, people got tired and homesick. As plastic tents fell over from the heavy snowfall, we began sharing big canvas tents. After four months of living there, I developed a routine. I'd take a water bottle to bed with me, and in the morning, after adding a few logs to the fire, I would boil that water to melt the ice on the spout of the water tank that had frozen overnight. Only then could we have access to the rest of the water. I had learned to always make the coffee first. (Like my friend Mark says, "Water protectors drink a lot of things, most of them ain't water.") Within minutes of lighting the stove, someone would come in and ask, "How long before the coffee is ready?" By the time it was done, there would be a line out the door. What can I say? You gotta be awake to fight a pipeline.

One dark night, the police decided to attack us with water cannons placed on the Backwater Bridge, something they hadn't done before. They showered us with freezing water and rubber bullets, as tear gas canisters flew above our heads. We walked back from Highway 1806 to the camp soaking wet, and several of us fell asleep near each other, sharing an inhaler. Several times, we woke up to one of us gasping for air. The chemical damage to our lungs from the tear gas had affected our ability to breathe.

Most of the time we assumed people in the rest of the world

had forgotten about us. "Are they still out there fighting that pipeline?" I imagined them asking. In the Tavaputs, I was used to the media silence; in Oceti Sakowin, it was deafening. In this silence, my mother prayed for me and my friends. One cold morning in December, I video-called her from the passenger seat of my friend's truck, crying.

"¿Qué pasó? What's wrong?" she asked.

The screen fogged up as I told her in one breath, "Mami, I'm going home. It's over. They're gonna build the pipeline. We couldn't stop them."

In February 2017 the U.S. Army Corps of Engineers was granted an easement to drill underneath the Missouri River, and it decided no further environmental analysis on the impact of the pipeline would be necessary. Law enforcement posted notices of eviction and began raiding the camp. On a television screen in Salt Lake City, I watched police violently arrest the people I had lived with for months as they destroyed the Oceti Sakowin Camp.

After I left North Dakota, my mom came from Perú to my house in Utah to visit me. After we had been separated for more than a decade and had spent the previous two years trying to get to know each other over the phone, her visa had been approved and we were ready to be together. A lot of hurt comes from family separation. A heart isn't an easy thing to fix; you can't just stitch it back together. Hearts are like mountains; they don't simply recover from destruction. A ravine had formed between me and my mother.

We made plans to go to Moab while she was in the States. I wanted to show her the things that kept me in the Southwest, I wanted her to understand me. Before leaving Salt Lake, we walked to the water well down Sixth Street and filled up a big bottle of water together. We walked to Liberty Park, and she listened to me talk about herbal medicine, police brutality, and life on the streets. Surrounded by the Wasatch Mountains, she sighed. "This place reminds me a lot of the Andes. It's cold, but not too much, and the sky is so blue." She paused. "I can see why you fight so hard for this land. It's your home. Your grandmother grew up in the Andes and so did your father, and now here you are in a different mountain range, in a different valley." She looked at me and frowned. "I wish you would stay home instead of going to these camps. When you got DACA, I thought that was it, I thought you would be safe after all these years of surviving. But instead, you keep finding new battles, ¿y para qué? For whose victory do you risk everything?"

The next morning before we headed to Arches National Park, she made me papaya juice and coffee. We ate breakfast together, and I bit into a piece of choclo she had brought from Lima. "It's my choice," I said between bites. "You asked me why I keep risking everything. It's because it's my choice, and it's worth it. When I was homeless, it wasn't my choice, but this time it's me making that decision." I looked her in the eyes and added: "I don't want to live in a world where I have to wait to be free. I want us to live free every single day of our lives, with citizenship or not. I want our people to be free, I want our relatives to be free. These borders sliced up the land and our communities and tried to strip us of our identity and culture, but this land is worth the fight. And our land is too."

Later that day we drove to Arches. Her eyes widened in awe as she stood underneath Delicate Arch. I watched her hike up the red sand and smile with triumph when she reached the top. The massive sand dunes towered above us as the sky turned into infinite shades of yellow, then red, then purple. Red rocks filled the distance as far as we could see; no matter which direction we would go, it was a comforting feeling holding up my hand and seeing the color of my skin on the earth.

A few weeks later, we were back at the airport hugging each other goodbye, just as we had more than a decade before when I left Lima. I watched her disappear through the gate and started crying. I closed my eyes and imagined her plane flying over the Wasatch Mountains, past the Book Cliffs, over the Tavaputs Plateau and the aspens, past the unfinished tar sands mine, over Delicate Arch, past the borders keeping me here. In my mind I traced a line alongside the long mountain range that connects North and South America, from the Yukon Mountain range all the way to the Andes, and to her, de montaña a montaña, mountain to mountain.

Mariella Mendoza is a multidisciplinary artist and media strategist with roots in the Andes and the Amazon rainforest. Mariella's work explores their personal experience as a queer undocumented migrant and has been featured in local settings in both printmaking and murals. Over the past eight years Mariella has worked strategizing for environmental justice alongside multiple local and national groups, including the Peoples Climate Movement, Another Gulf Is Possible, and Utah Tar Sands Resistance. Mariella is currently studying graphic design at Salt Lake Community College and is the Grassroots Action and Media Coordinator at Uplift, a youth-led, youth-powered organization for climate justice based in the greater Southwest.

Laurel Chen

POEMS

YOU SAY CITIZENSHIP I SAY A COUNTRY IS A CATASTROPHE

After Jasmine Reid

We live in a house quieter than our panic.
The front windows are tall and white. The glass is warped
from our desire.
We dress the windows with curtains but desire still floods
the house in summer.
I open the doors of the fridge when it is too warm with our
want.
I crawl inside there until my parents say time's up.
Want hums, warmer than wind or worship.
My father turns off the fridge one night when it sounds so
much like cicadas.
The house is so quiet, sometimes I can hear the metal pipes
rusting inside it.
My mother and I both cut our hair so that our bangs wave
like flags.

I paint my lips red and wear only darkened blues or greens.
My brother waits patiently at the mailbox, want bends his figure in half.
When the phone rings, I feel each note twist in the air.
Even sound is twisted under want.
If I reach out, I can almost touch it when it hums like that.
We hold our breath until our lungs vibrate like refrigerators.

Laurel Chen

auto-deportee aubade

alien anywhere is alien everywhere, alien
   cupping everyone i love softly
in their sleep.

   alien hushes my baby
cries at night when i think of my mother
   and the ways we refuse

to be loved.
FBI agent reading this right now
   i've been thinking about disappearing

for a long time anyways.
   i've made a playlist titled
wedding songs knowing damn well i can't

   actually get married because
someone has to leave first.
   this is the hard part, i think.

admitting this. one day we're drowning
   our legs in green bathwater
the next we're looking up

   one-way flights from SFO.
i'm waiting until someone self-deports
   to come clean. i'm waiting

for us to get green
cards before i get gay married.
morning comes and i dress

as the daughter my mother wants:
underwired, bare faced,
blackened hair, legs

closed. every cavity of my body
holds a secret: prescription pills,
lipstick stain, collarbone

tattoos. gender guillotined.
gender apology. dear mother, i'd slice
all the garlic if you'd believe me

when i say i love you. when i say
alien i mean i know we're not
meant to be here forever.

and i want that, too.
an excuse to dance
in public to a song that knows me

the way you don't, but love me all the same.
a love that feels like a skirt petering
around my ankles after a long winter

of strapping ourselves in
closer to our skins.
what we know is what we know

until it isn't. here's a bowl of bone broth and a jar
filled with wilting picked flowers.
loving means leaving

everything familiar behind, you
taught me this first. don't you remember?

Laurel Chen is a queer/trans/migrant writer from Taiwan. A fellow of Kundiman, Undocupoets, and Pink Door, they co-curate @shoutingpoems on Twitter. They are an abolitionist.

Lucy Rodriguez-Hanley

Guilty of Being Lucky

The charge nurse was about to scold me again. With a shake of her head, she came in and turned off the alarms that alerted her to a spike in my blood pressure for the third time that morning. My blood pressure was taken automatically by a machine every three hours, even while I slept or tried to sleep. I was twenty-seven weeks pregnant with my second child and on day five of hospital bed rest. It was supposed to be a night or two, but the doctor had yet to find a medication to manage my blood pressure.

"You can't keep watching that, you're going to give yourself a stroke," the nurse said, turning off CNN.

A young migrant mother was crying on national television over the kidnapping of her baby. Her eighteen-month-old was one of the hundreds of children ripped from their families during the third week of the Trump Administration's "Zero Tolerance" policy.

I had never been away from my two-year-old daughter until now. Our bedtime routine was listening to Stevie Wonder while I inhaled her baby scent. We whispered "I love you" back and forth until we fell asleep. Hudson is my miracle baby, conceived after five rounds of in vitro fertilization.

Matt and I were fortunate to have a baby to take home after so many losses.

I was hospitalized the night after her second birthday, separated for medical reasons—give my son a chance at life and to save mine. I felt guilty missing my daughter, knowing there were mothers out there who had no idea where their babies were. Hudson was safe with her father in our home in Long Beach, a mere twenty-five miles away from the hospital. We were lucky. I didn't dare complain, not even in private. One never knew when their luck might run out.

At night, in my dreams, I heard Hudson call for me, her cries mingling with those of the migrant mother calling for her baby. Then, my mother appeared. Her ghost sat in the visitor's chair. She'd died four years earlier, but it was her younger self who was there with me—the young Mami who left me and my three siblings behind when she and Papi migrated to Los Estados Unidos from the Dominican Republic. My mother had sat me down the night before they left and told me that she and Papi were going to Nueva York. Once they had jobs and an apartment, they would send for my siblings and me. The next day, Tía Vichy took us to a movie theater. We were excited to see *La Niña de la Mochila Azul.* But right before the movie started, Tía whispered, "Ya se fueron tus padres." A lump formed in my throat; my parents were gone. Would I ever see them again?

In my sleep, all our cries—mine, Mami's, the migrant mother's—merged into a symphony of sorrow.

The television was removed from my room. I stayed away from news on social media per doctor's order. Instead, I was allowed to spend one hour on the balcony. Untethered from the machines that measured my blood pressure and my son's heartbeat, the only beeps I heard were the car horns on Wilshire Boulevard. Suspended over the city, I allowed myself to cry. Why does having another baby and wanting to give my daughter a sibling come at such a high price? Why are the mothers who sacrifice and risk everything to seek asylum in the land of the free losing the babies they want to protect? Why is suffering so deeply etched in motherhood?

I kept Hudson's orange dinosaur and the tea set with the red cups and blue plates on the window seat near my bed. It made me feel close to her when she was not with me. I FaceTimed her throughout the day. She always leaned in close, kissed the screen and said "Mommy?" like she wasn't sure how her mommy got inside the iPad. She hugged the screen, kissed it again, and went back to whatever she had been doing. The migrant children did not have that luxury. Did they say goodbye to the parents they were taken from? Did anyone soothe them when they cried?

After five days of living in a Labor & Delivery room, I worried I might be forced to stay until I gave birth. My blood pressure was unpredictable, and my doctor thought I was about to develop HELLP or preeclampsia, two life-threatening pregnancy disorders. If we made it to full term, I would be away from Hudson for almost three months.

On day six, Matt brought her for her second visit. I heard her outside my room saying, "Hi, I Hudson," to everyone she came in contact with. She burst into my room and ran up to my bed. Matt took off her shoes and sat her next to me.

"Mommy, you're here," she said as she grabbed my face with her little hands, looked into my eyes, and slobbered me with kisses, setting off the alarms. I kissed her head; her baby smell was replaced by the floral scent of her baby shampoo. I held her in my arms and marveled at her perfect little face, just like I had when she was born.

At seeing her, all the bottled-up emotions skyrocketed along with my blood pressure. My doctor was concerned and suggested we limit her visits. How could he keep me from my daughter? Hudson came back three days later, that was the schedule Matt and I came up with so her routine would not get disrupted. But after her visit, my blood pressure stayed high the rest of the day. Thinking of my mortality and my unborn child's well-being, I finally relented and agreed to weekly visits, as the doctor had suggested.

Whenever I drifted off to sleep, I woke up to loud beeps from the alarms my elevated blood pressure had set off. Each time, my mother was in the chair, crying. It scared me to see Mami so vulnerable. I was afraid to talk to her. Did her ghost know something I didn't? Was I going to die in the hospital and leave Hudson motherless?

Una vez un ruiseñor . . . That night, Mami's ghost hummed her favorite lullaby from when we were little. She wore jeans and the pink-and-white crocheted sweater she had on when she picked up Ona and me at LaGuardia Airport. Her trademark brown pixie was a cascade of shoulder-length curls.

"Mami, am I dying?" I finally got the nerve to ask the apparition. She petted my hair like she had when I was a little girl.

"No, te estoy protegiendo," she said, kissing my forehead.

When she was alive, I did not have the courage to ask why she and Papi left the way they did. We rarely talked about the time we spent apart. Three months after they left the Dominican Republic, they sent for my brother and baby sister. Three months later, my sister and I followed.

The stewardess held my and Ona's hands as we walked to the luggage carousel. Papi was waiting for us outside the glass doors. Mami stood a few feet away holding Sonja's and Romulo's hands, afraid to let go for fear they would get lost in the crowd. We ran to Papi, he scooped us up and took us to Mami. Papi took my younger siblings so Mami could have her turn holding us. I held on to her longer than Papi. I breathed in Mami's new American scent, anxiety mixed with Ivory Soap and Jean Naté. My parents knelt and hugged the four of us. We huddled in a tight-knit cocoon, making up for the six months we were apart. Thirty-four years later, I finally understood why they hadn't said goodbye.

I delivered my son a day short of twenty-nine weeks, fifteen days after I was admitted to the hospital. The day before Charlie was born, President Trump signed an executive order to reverse family separations. Matt shared the news hoping to alleviate some of the anxiety I had over my son's premature birth. He weighed fewer than three pounds and was the size of a ruler. Between the drugs, the hormones, and not knowing whether my son was going to make it through the night, I felt like I had no control over anything in my life.

The day I was discharged, I had to leave my son behind.

But I pushed the guilt away, and I made Matt speed down the 405 to my daughter. I had been away from her for almost three weeks. I sat on the front steps waiting for her to come out of daycare. I had a C-section and couldn't pick her up. She ran to me and hugged me; she buried her nose in my neck, inhaling me the way I used to do with my own mother. I cried and she did too. "Mommy you're here," she said over and over between tears.

The next day, Matt and I drove to the Neonatal Intensive Care Unit to spend time with our son while Hudson was in daycare. On the drive home, we heard that a federal court had ordered the government to reunite the stolen children with their parents. They had fourteen days to unite families with children younger than five and thirty days to unite families with children older than five. I knew the trauma of separation was going to haunt these families.

Charlie spent seventy days in the NICU. I was devastated by his birth—the guilt of not carrying him to full term, being away from him, not knowing whether he was going to survive. But I never complained about the physical, mental, and emotional exhaustion. Every day, I drove three hours to and from the hospital, splitting my time between my kids and pumping around the clock to feed him. How could I feel sorry for myself when my daughter was healthy and my son was receiving top-of-the-line medical treatment?

By the time Charlie came home at the end of August he was like a healthy newborn, my blood pressure was back to normal, and Hudson no longer cried when I left the house. It took months to undo the trauma that our short separation caused her—a separation where she was safe at home with people who loved her. How will the trauma inflicted by the Trump Administration on thousands of children and their mothers ever be healed? Will these families ever recover the way mine did?

My son is now almost three. He is healthy and thriving. When I hold him, I often think of the migrant mother I saw on CNN, hoping she was one of the lucky ones who got her baby back.

Lucy Rodriguez-Hanley is a creative nonfiction writer, filmmaker, and mother of two. A Dominicana from Washington Heights, she is now living in Long Beach, California. Her memoir-in-progress explores the effects that migration, assimilation, and maternal rage have on the narrator's life. She is a strong advocate for representation of women and BIPOC in creative spaces. She is the chapter liaison for Women Who Submit and is cofounder of the Long Beach Literary Arts Center where she co-leads the Long Beach chapter of Women Who Submit.

Emilia Fiallo

All the Little Pieces

You will never be completely at home again, because part of your heart always will be elsewhere. That is the price you pay for the richness of loving and knowing people in more than one place.

—**Miriam Adeney, *Kingdom Without Borders* (2009)**

I

It's a late July summer morning. One of those humid mornings you feel on your skin, when even noise seems to carry heat. In our apartment, David and I pick up our binder, a two-year curated collection of our time together, serving as proof of our legitimacy as husband and wife. We have rehearsed our story and memorized senseless details about our lives—phone numbers and past addresses, social security numbers and dates. Once in the car, I quiz David on my old addresses. He gets the first one wrong. My anxiety is intensified by the day's heat, and the air conditioner, though blowing at high speed, offers no relief.

It is 2707 East 65th Street, David. Mill Basin, remember?

He doesn't remember the address, but I know he remembers the stories behind the pictures I have shown him of the fourteen years I lived in Mill Basin with my family. My older sister hugging my dad in front of our Christmas tree. Me as a teen with over-plucked eyebrows and side bangs. My sister holding our pet rabbit against her chest, smiling big. Pictures of us around birthday cakes and Mom's best meals decorating the table. Always the four of us. Always home.

II

In 1998, we left our first home in Ecuador. I was eight years old; my sister, Mia, just a year older. My mom, forty-one. We came to reunite with my father in Brooklyn. He had left a year before us and soon realized he could not build home without his girls. He sent for us and we reunited. After six months, our tourist visa expired, and just like that, we became undocumented.

Though we began new lives in the United States, my family did not lose hope of going back to Ecuador. Our hearts were still in the home my father had bought for my mother when they married in the 1980s. Our hearts remained in the rooms we had left behind with toys and our twin beds. And in the front yard adorned with roses and avocado and orange trees that gave all year round. Our hearts clung onto the memory of relatives who called to ask the same question, *And when will you all come back?* We knew there would be a time when we would return. This was a simple idea and a mutual understanding among our family for a long time. Whenever life got difficult for us with an unfair landlord, rumors of more frequent ICE raids, or when we just felt homesick, we would always say, *One day we'll go back home.*

Things got complicated during my egocentric teen years.

At fifteen, I started to reproach my father for bringing us to this country without a plan. His only response was that he wanted to give us a better life. But the answer never satisfied me. Living undocumented was a better life? When I failed to get into college despite decent grades, I blamed him. When I couldn't find jobs because I did not have a social security number, I blamed him. The times I couldn't travel, I blamed him. The car I couldn't drive, I blamed him. I aimed all my undocumented emotional bullets at my father. Because I could not hurt this country, I hurt him.

Once I grew older, I would eventually come to regret the pain I'd caused him. Life often teaches children, the hard way, to venerate the struggles they once criticized their parents for. Unfortunately, my father would not be beside me when I came to such a realization.

In 2015, at sixty-seven years old and after twelve years of working in a pen manufacturing factory, my father decided it was time to go home to Ecuador. Despite having a legal social security number, something he managed to get in the 1970s on one of his trips to California, he learned he would never be able to retire and enjoy his pension because he did not have an actual green card. My dad grew older and more fatigued. His purpose as a migrant dad now fulfilled—seeing both of his girls graduate from college after years of undocudrama, of never having enough for tuition, and of my mother working endlessly to support our college dreams—my sister and I obliged and booked him a one-way ticket back to Ecuador. We sat around the laptop selecting his seat number, while my dad held on to his passport and watched us click around a website he did not understand. My father left in July 2015. I heard an old friend call my dad's decision a sort of "self-deportation." Others corrected this and called it a "voluntary departure." Family in Ecuador called it coming home. I knew it as the day my dad left home.

My father's departure hurt me.

My mother's departure changed me.

At sixty years old, once she was done raising all her children, including the ones she cared for as a babysitter for an Italian-American family, my mother decided to leave as well. The babies grew and there was no more work for her. Under DACA status, my sister was now working full time as a paralegal, and I had a job as a teacher lined up for the fall. Alone at home most of her days and without employment, my mother decided it was time to go, with faith that we were grown up enough to live on our own.

When she left, she tried to take it all—her best serving plates, tableware, appliances, clothes, and her comforters. She even took Rubi, our family dog. If she could have, she would have packed her daughters in a suitcase, too.

Home at 2707 East 65th Street, without them, was no more.

III

I am nervous, and I take it out on David. The traffic looms ahead, and he has a tendency to speed, which adds to my nerves. We both get quiet. After a while he puts his hand on my knee, and I realize he is nervous, too. He has never felt the pressure of having someone else's life and dreams depend on him as much as at this moment. I wish spouses didn't have to be the ones to do this.

At first, I was sure that the powerful undocumented youth movement catapulted by the DREAM Act and intensified by the endless deportations would give us all the path we needed to legalize ourselves and our families. If not that, then I was sure that my work as a student and later as a teacher would eventually prove me "worthy" of papers. If only I worked hard, if only I showed myself to be *useful* and *essential*, this country would create a path for me and others like me. But in the end, none of this moved this country to turn around and look at me.

David and I met on Instagram. We were always embarrassed to tell people about how we met. Now we laugh about it and call it fate. I was twenty-seven years old, mourning my father's departure and preparing for my mother's soon-to-be self-deportation. David was twenty-three years old with a permanent smile on his face, driving way too fast as he shared stories about his beloved birthplace—the Bronx.

I told him my migration story and shared what was happening at home, every step that had led to my sister's and my decision to let our parents return to Ecuador, how they were getting old and with Trump's victory we knew things would get worse. We persuaded them we'd be okay without them. We didn't want to give our parents' golden years away to this country. I told David about the anger I felt at having to choose between the life my parents built for me here or life with them back in Ecuador. I shared my frustration at how my father's departure should've prepared me for my mom's decision, but it had only made me more afraid.

The first time David met my mother, he was nervous and so was she, but my mom liked him. She's always favored boys. He wished her a safe trip back home. My mom asked him to keep me safe. David has kept his word.

David, what about the address after that one?

1225 Avenue R.

Good.

The first apartment my sister and I lived in without our parents was in a building where everything felt cold. In this apartment on Avenue R, I witnessed my sister cry. It was the first time I felt my sister, always composed, lose control of the rage she must've held on to for too long. In this apartment, we drifted away instead of staying together. That apartment was the incubator of our sorrow, bursting from years of sister rivalry, the pain of losing both parents, and the thousand personal issues we kept from but took out on each other.

My sister. We failed to make a home by ourselves without our parents.

I would go to a bakery and buy four cupcakes instead of two. I would call my parents and tell them, *I'll see you when I get home*, and then I'd correct myself and tell them I would *speak* to them later. I would dream of the four of us together and cry when I awoke to find myself alone in that apartment. Sometimes, I'd hear noises in the kitchen and feel my mom there. One night I got drunk while out with David, and I convinced myself I still lived with my mom. I demanded he take me home—to my real home on East 65th Street. David grew concerned. I stopped cooking. I stopped eating at home. The kitchen grew vacant. Food rotted in the fridge. I accidentally broke one of the mugs my mom left for us, and I felt so much guilt over breaking one of the few things she had left behind. Keeping her things intact kept her near me.

That year, Mia and I both learned why my father had brought us here. It wasn't necessarily for a better life. There was no guarantee of that. It was my parents' love. I understood why my dad couldn't survive this country without us. Home wasn't home if we weren't together. Still, my sister and I went our separate ways to give each other the freedom to search for independence and solitude. When I moved into my own apartment in Queens, far away from everything that reminded me of home, I felt relief.

While I mourned for my parents' absence, David patiently loved me and gave me hope of a new home. He moved into my apartment in Queens shortly after I moved in.

IV

David and I make it through the maze of the building where we are to have our first interview with the immigration officer. We sit at the waiting area trying to distract ourselves with the TV—taking in a few moments of screen and glancing back at each other for reassurance we are still here. No more quizzes. Just silence.

I notice the fast tapping of his left foot and the nervous caressing of his beard. I think of the possibility of losing home again—the home with David, my husband. I think of the babies we talk about having, but which I hold back on because I can't imagine pregnancy and motherhood without my mother by my side. The little dog (for me) and big dog (for him) we want to rescue. Our Sunday morning routine of pancakes and a sitcom. The home we had been building for the past two years can be taken away today. The day we got married, a year ago, my father did not walk me down the aisle. Instead, I walked alone. The photographer snarkily asked where my family was, as if it were inconceivable that the bride would have only three people at her wedding. In the middle of getting my makeup done, I ran to hide in the bathroom, squeezing the tears back into my eyes. Seeing David at the end of the aisle made me feel safe.

David initiated the process of adjusting my status because my sadness over missing my parents permeated our home, especially on hard teaching days and holidays that suffocated me. When I ask David why he wanted me to get my papers, his answer is always the same—*Because I want you to be happy, babe.*

We are finally called in, and I let David speak first. I am too scared. He and the immigration officer quickly fall into easy conversation, so my anxiety begins to loosen its grip. We hand over papers and documents, shared bank accounts with the little money we've saved over our time being married, taxes done together, the lease of our studio apartment in Queens, and the countless other receipts of a life together.

The immigration officer asks to see pictures of our wedding. He asks about our trips. I remember the one to Puerto Rico where David fell off the hammock, and another day when we almost got stranded on some hill with no cell service. He asks about my parents as he looks at the pictures of David meeting them in Ecuador to ask for my hand in marriage. He'd gone to see my parents and took his family as well, since being undocumented prevented me from being there to introduce him to my father and extended family. The immigration officer asks about our future. He never asks about addresses.

When we leave the building around noon, David and I step right into the peak of the day's heat, but it doesn't bother us. I want to see fireworks around me. I want the sky to break open and Jesus Christ himself to come down with that precious green card. I want it to start raining so that I can feel myself in an R&B video, unable to distinguish raindrops from my own tears. But I don't cry. I want to but can't. I want to call everybody and not tell anyone at the same time. I call my parents.

V

My green card arrives in the mail six months later, when I am away in Florida at a teachers' conference. When David calls me to give me the news, the thought of holding the one paper that will bridge the distance to my parents makes me feel like everything is possible now. But I think about my sister, and the feeling of never being truly complete comes over me again. My sister is still undocumented—her whole career and livelihood in limbo, a feeling too familiar to her at this point. Still, when I tell her my good news, like a little girl she jumps to embrace me with excitement, *You're going to see them again!*

When I go home after twenty years, I watch my parents

picking fruit from their avocado and orange trees, watering their roses, playing with their three dogs—all rescues, all girls. After a day out in the yard, they go back inside for café, and I join them. Home is here with them. Home is also in Queens with David. Home is in Brooklyn with my sister, too. This is what being an immigrant does to you—it splits your heart into little pieces to be housed in separate homes where everyone misses you whole.

Emilia Fiallo is an English language arts high school teacher and continues to support undocumented students and immigrant families across NYC public schools. David and Emilia are celebrating three years of marriage. They are in the process of adopting a dog. She has not seen her parents since COVID hit but occasionally visits her sister in Brooklyn.

Julio Salgado

A Moment for Two (2021)

Julio Salgado is a visual artist who happens to be undocumented and queer. Being undocumented and queer has fueled the contents of his visual art, which depicts key individuals and moments of the DREAM Act and the migrant rights movement. Undocumented students, organizers, and allies across the country have used Salgado's artwork to call attention to the migrant rights movement. Salgado is the co-creator of the Disruptors Fellowship, an inaugural fellowship from the Center for Cultural Power for emerging television writers of color who identify as trans and/or nonbinary, disabled, or undocumented/formerly undocumented immigrants. His work has been displayed at the Oakland Museum, the San Francisco Museum of Modern Art, and the Smithsonian.

Francisco Aviles Pino

POEMS

Fruit

After Lucille Clifton

The branches / don't hold their fruit too
long / but look: my fingers raised / to the
moon / after /gathering soil /knowing yours
are pointed / too / & all I know of /
freedom / is that / it's about walls / gates /
falling / to gardens / that aren't / secrets. /
There's a reason / a tree bears fruit /
teaching / flight / testing gravity / and
making friends / with the fall

Unite Here Local 11

After Ross Gay

Look, my hands are made of other hands that were made of other hands & those hands made car parts for Ford, my sister's hands cooked the dorm food at UCLA, the pizza and veggie patties that I bring home, my hands make their Hyatt beds, and they sometimes leave no tip when they check out, I always wonder why, but I also never dwell. But I will not surrender, I am good at what I do and I love the people I work with so much that I sometimes meet after work with them to talk about justice. I imagine more than this, I imagine my son at school, getting better at dividing. My son attends Sunkist Elementary, he walks there and back home alone, the way I taught him to, focused and protected by prayer. My daughter studies sociology, she takes the 42 bus to UC Irvine and sometimes forgets to call me when she's back home with her brother. My color is pink, my room a collection of long breaths. My mirror, a reminder of my Mom, my clothes, and all the comfort they bring, and the bed where I hold all that I am tightly.

For William Camargo

An ode to journalists

After Safia Elhilo

It is 4 a.m. on Monday morning in Acapulco, Guerrero, México, & before he is my grandpa, Rosario is packing newspapers on the back of his truck with the help of his daughter Paloma. They drive down the colonia at 5:30 a.m. sharp, & before she is my mother, Paloma, the oldest of 8 kids yells into the megaphone the news of the day,

encontraron a 6 periodistas muertos y
siguen 8 desaparecidos.

A country that is no one's, a country called México was no longer safe,
so Paloma ran away from one crumbling house to the other but needing to send her two kids first—And I was lucky to not find a camera trying to photograph me in a hotel,
dirty and alone with my sister, before journalism, I knew that I would remember the smells and that this story would always matter & I believe this to be the reason why my mom never says goodbye in phone calls now, & how after one call during her first trip north, we had to pay to talk to her after that.

Over ten years later,
after the migration attempts, in between the separation and some assimilation,

my mother asks me to change the channel when the news comes on Univision.

After I told my grandpa that I want to be a journalist, he was silent for almost two minutes and finally said, *ten mucho cuidado.*

Francisco Aviles Pino is an undocumented queer writer and producer based in Los Angeles, California, whose work focuses on incarceration, migration, and culture. They have published multimedia content for the ACLU, The Intercept, *Vogue*, *OC Weekly*, and Brave New Films and have consulted for winning political campaigns and candidates. Aviles Pino is an alum of the Macondo Writers Workshop, the NALAC Leadership Institute, the Poetry Foundation's Emerging Writers Fellowship, and the NCCEP-GUALA Reach Higher Initiative during the Obama Administration. They worked as a Staff Community Organizer for the Orange County Congregation Community for five years, where they supported numerous local and statewide campaigns and organized several intergenerational coalitions. Born in Acapulco, Guerrero, México, and raised in Los Angeles and Anaheim, Francisco is an artist who is always thinking and working through the intersections and conflicts between history, documentary, theater, participatory research, poetry, and public policy.

Angel Sutjipto

Discretion

> In telling you all of this in this way, I am resigning myself and you to the idea that parts of my telling are confounding. I care about you understanding, but I care more about concealing parts of myself from you.
>
> —**Eve Tuck and C. Ree, *A Glossary of Haunting* (2013)**

August 18, 2014—Hunched over a plastic chair in secondary inspection at JFK Airport, elbows pressing into your thighs, backpack weighing down your shoulders, you hold your military green Indonesian passport in your right hand. In your left, you hold the original copy of your DACA and Advance Parole documents. After spending ten days with your white colleagues in Kampala, all you want is to go home—to the studio apartment in Forest Hills that belongs to neither you nor your mother, but to her husband and his anger. Home has never been a safe place for you; yet you find yourself missing your own bed.

Your lawyer's warning swirls in your mind as you reread the bolded text of your immigration documents: **"Parole into the United States is not guaranteed. . . .** [T]he Department of Homeland Security retains discretion to deny you parole if the Department determines approving your parole application would not serve the public interest." You do not need to read between the lines to recognize the threat. "Discretion" means that the US government can deny your request for reentry for any reason. "Discretion" implies that the government can upend the life that you and your mother have spent a decade building in a matter of seconds if it believes that your presence will not serve its interests or is a threat to national security. All decisions final—there will be no recourse.

(Years later, you will finally understand the message behind DACA and Advance Parole, both of which are exercises of prosecutorial discretion by the Executive Branch. Discretion: The US government can establish arbitrary age limits and guidelines to determine who is considered a human being. You have been dehumanized for so long that you glossed over this form of control and violence. One that denies you your full humanity—the right to make mistakes and to redeem one's self without the threat of incarceration or exile. One that drives a wedge between you and your fellow im/migrants because DACA comes at the expense of other im/migrants with prior criminal convictions.)

Sitting in the third row, closest to the exit, you strain yourself waiting for someone to mispronounce your name. But all you can hear is the hum of government bureaucracy masking the stillness of lives stalled. A few seats away from you is a South Asian family of three—the father rolling and unrolling a newspaper, their young daughter lying across her mother's lap. Your eyes dart to a group of Chinese international students sitting in the front row, then to the Black and Brown men scattered throughout the room, before landing on an elderly East Asian woman who's lying down in the last row. All of us silenced by our perceived powerlessness.

(Later, you will ask yourself: "Why didn't I try to speak with any of them? Didn't I call myself an organizer? Why did I choose alienation instead of connection?")

For all your bravado—insisting to the gate agent at Kampala Airport that your Advance Parole document meant that the US government would likely allow you to reenter, to telling your colleagues that they need not wait for you at the luggage carousel—the truth is you are Still Afraid of possibly being interrogated, detained, or deported.

Last year, your boss and mentor invited you to go to Phnom Penh to witness proceedings in the Extraordinary Chambers in the Courts of Cambodia. But the words "I can't" tumbled out of your mouth before you could stop yourself. You came to this work—genocide and mass atrocity prevention—because you grew up undocumented and you needed to understand how people could be so cruel to each other and to believe in redemption.

(Discretion: Redeem yourself by excelling through your bootstraps and embracing settler-colonialism and anti-Blackness.)

Three months ago, when your boss asked if you wanted to join your colleagues to conduct human rights fieldwork on the LGBTQIA+ community in Kampala who are fighting back against Uganda's anti-gay legislation, you said, "Yes, I do." You had spent the past year weighing the risk and talking to two DACA recipients who had traveled on Advance Parole. Both reentered the US without much fuss. But fear and anxiety have controlled your life for so long that you have mistaken them for friends; nothing can turn off the catastrophizing refrain that loops inside your head.

(Discretion: If your skin color, disability, complex trauma, or poverty makes you a target for law enforcement, the US government will say that you brought it upon yourself. The government's immigration attorney will say, "We have every right to remove you in the name of public interest. We are initiating deportation proceedings against you because you are a threat to society." If you have learned anything from working in the human rights field, it is the irony of appealing to nation-states for one's humanity when they are the perpetrators of violence. As a survivor of domestic violence, you know that asking your abuser to stop is ineffective, that abuse leaves indelible marks on your psyche, and that leaving feels physically and emotionally impossible. You will ask yourself, "Will the US ever break its cycle of abuse? Will any of us ever be protected from this country's violent outbursts?")

Now, sitting in secondary inspection, you wonder whether your relative privileges—lack of a criminal record, eligibility for DACA, and a university job—will be enough to protect you. You know that there is not much that can protect Black and Brown bodies ~~in~~ from this country. When news of Michael Brown Jr.'s murder at the hands of Ferguson police reached Kampala, one of your white colleagues wondered aloud whether "rioting" could effect positive change. The invisible chasm you have always felt toward your colleagues, as the only queer person of color on the team, suddenly feels impossible to bridge.

(Next year, one of your supervisors will remark that she is glad to work with asylum seekers because they're angels. The words left unspoken will linger in the air: Because im/migrants with criminal convictions

deserve to be deported; because defending im/migrants with criminal convictions takes up too many resources and too much time. You will begin to suspect that, although your colleagues came to this work with good intentions, they had neither confronted their own whiteness nor scrutinized their own intentions. The projects that you manage will begin to reek of neocolonialism and White Savior–ism; you will know then that it is time for you to leave the human rights field.)

Time coagulates in the room. You worry the edges of your military green Indonesian passport with the oil and sweat that have accumulated on your fingertips. As you reach into your pocket to check the time, a CBP officer catches your movement and barks, "No cell phones allowed!" Glancing around, you cannot seem to find a clock anywhere in the room. Perhaps there are no clocks in secondary inspection—or rather, what use is knowing the time for a group of people existing in a suspended state?

("Control people's time and you control their reality," one of your mentors will say. For those of us who are targeted by the State—Black bodies, Indigenous bodies, migrant bodies, queer bodies, disabled bodies—we have a complicated relationship to time. Stolen time, serving time. Running out of time: to find a job to pay for the lawyer, to file the application, to enroll one's children in school, to schedule a doctor's appointment, to file an appeal. Working overtime. Slowly killing ourselves only to be erased by time. Is it any wonder that we are time travelers?)

In 2003, shortly after settling on the land of the Matinecock Nation without their consent nor the permission of the US government, your mother purchased phone cards from a window booth in Flushing to call your father in Jakarta to let him know where you are. Your mother had nothing to say to your father, so she passed you the phone. Hearing your father's voice—"Apa kabar, Angel?" he asked—disoriented you. You fell silent, unwilling to voice the thoughts in your head: "How am I? I'm not sure I'm real, Papi. That's not what you want to hear, I know. How am I?"

(The next time you speak to your father will be in seventeen years. Even after all that time, you will still struggle to forgive him for the way he treated your mother.)

You lose track of time while time traveling. The sound of footsteps brings you back into the present. Your eyes track a CBP officer carrying a juice box and a bag of chips walking toward the elderly East Asian woman who has curled into herself to stay warm. Her white hair, like a puff of cloud, reminds you of your grandmother and open skies. You think to yourself, "How long has she been here that they have to give her food?"

Finally, your name is mispronounced. You grab your backpack, your passport, your immigration documents, and walk to the front of the room. You look up at the CBP officer seated on the elevated dais as your mind races through a list of acceptable answers.

He gathers all of your paperwork and holds up your passport to confirm your facial features. "Do you have anyone who can petition for you to adjust your immigration status?"

"No," you reply.

(There will be no way for you to adjust your status except through marriage. But having grown up witnessing domestic violence, you have ruled out marrying for papers long ago. You would rather remain undocumented than to allow another person to have power over you.)

The officer types into his computer, then stamps your passport: paroled into the United States until August 17, 2015, for adjustment of status. As you walk toward the exit, adrenaline pounding in your ears, you look back and spot a familiar puff of cloud. You realize that the elderly East Asian woman is still lying down in the back row.

(Five years later, you will realize that your responses to the CBP officer were not important. What mattered was that the threat had been conveyed. Discretion means: Behave. Make yourself small and keep to yourself. We control time and, therefore, your reality. We can choose to inflict fear and dread indefinitely—or we can release you. Discretion: All decisions final.)

Angel Sutjipto was born and raised in Jakarta and Kuala Lumpur. For the past eighteen years, they have resided on Lenape and Matinecock lands (aka New York City). Their creative nonfiction essay "dis place" was selected by Alicia Elliot as runner-up for *Briarpatch* magazine's 2018 Writing in the Margins competition. They are an alum of Voices of Our Nation (VONA) and the CUNY Baccalaureate for Unique and Interdisciplinary Studies. Previously, they have organized with ICE-FREE NYC and RAISE (Revolutionizing Asian American Immigrant Stories on the East Coast), where they cofounded AMPLIFY(HER), a zine by and for undocumented womxn from the Asian diaspora. Their literary ancestors include Gloria Anzaldúa, bell hooks, Audre Lorde, Cherríe Moraga, and Arundhati Roy, among many others. In their spare time, they sing, tend to their plants, read the Tarot, and dream of working and writing on a farm one day.

T. Lê

POEMS

I unravel to tether myself

two eyes *bleak* *obsidian*
following

THIS WAY
beneath the flickering lights *white cast* *halo*

savior the limbs echoed
against the current of bodies
savior the limbs echoed
against the lapping goodbyes

forwards and backwards
sway the blue seats
beneath flickering lights *soft yellow* *halo*

savior the limbs scratched
against two small persimmons
each to fill an hour *soft bruised*
flesh

like *soft bruised*
body flying   *releasing*
        limbs into the horizon
        beneath the flickering blue void

you insist on my native tongue

and I think of fish mint and its facetious echo
as it rolls off my tongue   a façade

of an ocean growing alongside my father's
picket fence   I think of survival   I think of

the necessity to be near water, and with it, the many
faces breaking through the viscous names of *bác* and

dì   undulating   waiting   to be called
          I think of their voices struggling for clarity

a codex   stolen   for   joy   grief
and family   I think of my younger self

at 3   blue dress   stone steps
more promises of   *last picture before we go*

          I think of families   sleeping on a small boat
swaying to the ticking engine

          I think about the shortage of air   I think
about the missing faces until I expand into silence

          I think about the chirr my throat makes when it
marks a blunder in my voice   *I'm sorry*

my parents don't speak English   I think about
the clicks of rice paper   sharp and brittle

but softened by water   I think about my mother's
hands kneading grief into each of her recipes

to blanch our sins of leaving a country   I think about
my face   and   I think about the bitterness of not
looking like you

NOTHING IN PARTICULAR

I write about creaky wooden floors. But mostly, I
like to write about the bare feet that walks them.

It's the coolness against a living body that snags me. Maybe
it's the living body. *Wholesome.* The hard wooden floors.
The bare feet.

The body, *belong.*

I write about the thistle of photographs, *hidden* or *partly forgotten* resting
on the closest shelf. I know the faces are no longer there.

My father calls them *gardenias.* Tiny faces drowning in a film
of dust, *hidden* or *partly forgotten.* I write about their drooping faces

in white chiffon collars (the dead must also look nice). I write about
Uncertainty and she wears a face drowning in white chiffon collars. She mumbles

too much, and her eyes are a bit slanted. I write about her jaw not wanting
to crack from the stress on one syllable. *English.*

I write about the color white, for its coolness in drowning tiny faces
wearing white chiffon collars. *Absurdity.*

I write *American.*

I write *space.*

I write *you must take what is not yours.*

I write about a conical hat, turned upside down, filled with congee. It fed
five villages. And in turn, the five villages planted green mangoes.

When *Absurdity* visits, they offer *Absurdity* green mangoes.

Absurdity purchased the conical hat at a souvenir shop for twenty-five cents.

Absurdity.

I write about *Absurdity* tracing mouths belonging to tiny faces of *Uncertainty.*
Absurdity stealing words for stories already written.

The already written stories are splattered in maroon and yellow. They

paint the bark of my grandfather's guava trees. *Ông nội*
  or is it

Ông ngoại? I've forgotten.
His stories, I've also forgotten.

So, I write about *Nothing.*

Absurdity cannot take *Nothing* if there is *Nothing* to take.

T. Lê is a Vietnamese American artist. She immigrated to the US in the early 1990s with her ba, Sinh, her mẹ, Vân, and her em gái, Thảo. She has a fondness for words, both in poetry and in acting. She loves all work by Mary Oliver. On occasion, she'll bask in a cup of bad coffee. Currently, she is working on her first poetry collection.

Elías Roldán

A Permanent Stitch

Made from a radiant polyester satin fabric in red, the color of passion, embroidered with the Greek key on the hem and decorated with gold-threaded streamers and sequined stripes that would catch the stage lights, the costume was a vision to behold. A flamenco cut rather than traditional to give the dress more flare. At the performances, the dancers brought my creation to life as they twirled around the stage to the mariachi music.

I poured my heart into this Jalisco dress and all of the costumes I created for the ballet folklórico company Grandeza Mexicana when I became its costume designer in 2006. Our dream was to win the coveted Lester Horton Award, which recognizes excellence in the Los Angeles dance scene. No Mexican ballet folklórico company had ever won the award. Our dance director submitted entries for best choreography, music, and costume design categories. I was shocked when I heard the news that the costume design entry had won! I read and reread my name, thinking they had made a mistake. Competing with so many talented designers, I didn't feel I had what it took to take home the award. The day of the reception, I walked up to the podium to receive my award thinking it was a dream I would wake up from. I had just beaten out many talented designers, and it meant a new bar had been set for me professionally. As nervous as I was, I felt extremely proud to be representing my community and celebrating my Mexican culture through art.

Winning the award also boosted my confidence and allowed me to dream about the future. When the opportunity to open my own business came about, I jumped at it. I had been working as a costume designer for someone else, but when my boss decided to close the shop, the owner of the building encouraged me to open my own business and offered to transfer the lease to me. I had no knowledge about running a business or resources, but I had proven myself capable of winning awards and I felt it was worth the risk.

I had just enough money for the deposit and the first month's rent. I quickly realized the mess I had gotten myself into. With no new clients coming, I barely had enough money to cover the second month's rent and not enough for the utilities. Instead of sleeping, I read articles on business management or brainstormed ways to make money. I learned from some clients that I could register my business with the City of Los Angeles, obtain an IRS tax ID, and apply for a small business loan.

I didn't realize it was possible for an undocumented immigrant to be a business owner. When I showed up at city hall, my heart was beating fast, and my hands were sweating. At the awards ceremony, the worst-case scenario was going home empty-handed if I lost. Here, I was dealing with the government, and the process was unfamiliar to me. What if they asked me for my green card? Would they call the police or ICE? Would I be deported? To my relief, the process turned out to be a smooth one, and I walked out with the document bearing my new business name in black and white—ELIAS DESIGNS, INC.

With my newfound confidence and business license, I dressed in black slacks, a blue shirt, and black loafers and went to the nearest bank to apply for a small business loan. This was the last step I needed to get my business off the ground. As I sat in the lobby, I started dreaming about my thriving business: all the fabric and equipment I could purchase; all the beautiful designs I was going to create. I could see my business buzzing with clients, employees busy fulfilling orders, the sewing machines running at full speed.

"Excuse me, sir?" The loan officer's voice pulled me out of my reverie. "I will be assisting you with the loan application."

I followed him through the lobby to his cubicle. I presented my business documents and asked him how much I could apply for. After he reviewed my paperwork, he asked me for an ID and my social security number. I provided the Individual Taxpayer Identification Number ID I had been assigned by the IRS to pay taxes. I was not expecting what came next.

"I'm sorry, Mr. Roldán, but we're unable to issue a loan to individuals with ITINs. You need a social security number."

I was confused. I thought all I needed to apply for a loan was a registered business. I was now a business owner. I had an IRS tax ID. I paid taxes. Wasn't that enough? I felt as if all my dreams had been sewn with a temporary stitch and this loan officer had just pulled the end of the thread, unraveling them all. I walked out of the bank feeling ashamed for even thinking a bank might extend credit to someone like me.

I went home choked with anxiety. I still needed to pay rent and utilities, and to make matters worse, the first quarterly payment of my business taxes was due. No one had told me that I had to pay taxes even if my business wasn't making a profit. I had to figure out what I was going to do, fast. I realized the only way someone like me could make it was to work twice as hard. There were days when I would work nearly twenty-four hours straight. My social life disappeared. While my friends were going to parties, I was at my shop, designing and sewing anything that I was lucky enough to get paid for—wedding and quinceañera dresses; salsa, flamenco, and bachata costumes for dance competitions; and mariachi outfits. I posted photos of my creations on social media and did everything I could to reach new customers.

Every month, I made only enough money to pay the lease and purchase materials. I would fall behind on the utilities but catch up a few days before the electricity was cut off. I continued to receive quarterly statements for taxes overdue. I thought the IRS would know I wasn't making any profit and was barely staying afloat, and I didn't know I could make payment arrangements with the IRS, so the taxes I owed just continued to pile up.

One night, I was in my shop, working late as usual, applying the last touches to a Veracruz dress, when there was a knock on the door. I wondered who it could be. I wasn't expecting any customers. I hoped it was someone from a dance company looking for a new costume designer. I opened the door, and to my surprise, it was two sheriff's deputies. I started shaking.

"Are you the business owner?"

"Yes," I said.

"Can we come in?"

I barely had enough strength to open the door. I was petrified, thinking they had come to arrest me because I had not paid my taxes. They asked what my name was and said they were doing a routine check in the neighborhood. One of them walked around my shop, moving around bolts of cloth lined up against the wall, picking up spools of white ribbon and lace, and staring at the unfinished Nuevo León jacket hanging on the mannequin. The other one stood next to me with his hand on his holster. I felt that my world was about to crumble. I could see myself being handcuffed, taken to a station, and put on a bus back to México. Then all of a sudden, they thanked me and left.

A few days after that unexpected visit, I received a letter from the IRS requesting a list of my business inventory. Were they going to take action against me? The letter terrified me. Any mail from the government always sends me into panic mode, but that, combined with the sheriff's visit, was too much. The stress of running my business finally got to me, mentally and physically. I ended up in the hospital, adding medical bills to my expenses. As an undocumented immigrant, I didn't have access to health insurance, and I didn't qualify for Medicaid, yet I was expected to pay taxes. With bills piling up, I was no longer able to afford the lease of my shop. I needed to pay the IRS before I dug myself deeper into the hole. I gave up the shop and decided to continue to run my business from home.

On June 15, 2012, President Obama announced the DACA program. It would protect people like me, who came to this country in their youth, from deportation and offer work permits. I celebrated the news thinking I could dream again, but my excitement was short-lived. I did not meet the age criterion. I was forty-one, and the cutoff age was thirty-one. Even though I was ten years too old to be considered a "Dreamer," I wasn't too old to dream. I got back on my feet and reopened my shop. This time, I was more knowledgeable about how to manage a business. I was also more confident in my talent. I wanted to continue contributing to my community. I wanted to do my part in celebrating our beautiful Mexican dance and music traditions. Also, I didn't want to see myself as a victim but rather as a survivor capable of accomplishing my dreams regardless of my immigration status. I couldn't let that define me. When I see ELIAS DESIGNS, INC., sewn on one of my creations—a new mariachi outfit or dance costume—it reassures me that I've come a long way and have much more to give.

And sometimes, when I feel like giving up, magical things happen. This was the case in 2017 when, one day, I received a phone call.

"Mr. Roldán?"

"Yes?"

"I'm calling from Pixar Animation Studios. I'm one of the advisors of a movie we're about to release. The premiere is taking place in a few weeks in L.A., and Marcela Davison

Aviles highly recommended you. We're looking for someone to design outfits for our premiere red-carpet gala as well as the Grammys."

As it turns out, the movie the person was referring to is one of the most successful and beloved Pixar films, *Coco*. They asked me whether I was able to travel. I was nervous to get on a plane because of my immigration status, but I knew this was one of those once-in-a-lifetime opportunities. I said yes, but quickly added that I have a phobia of flying and asked whether I could instead drive to the Bay Area where Pixar Studios is located. Luckily, they were very accommodating.

At the Pixar security desk, when I was asked for an ID, my hands shook when I handed them my Mexican passport. They simply glanced at it and let me through without a problem. Moments like these bring up my insecurities, but I pushed them away and focused on what I was there to do.

When the movie events took place, and I saw the photos of the premiere, the Grammys, the media interviews here and abroad, I felt so honored and proud to see my designs on display. Pixar director Lee Unkrich wore a three-piece cream-colored suit with a skull design inspired by the charro outfit from the movie. Marcela Davison Aviles looked elegant in the china poblana dress I had finished just a couple of days earlier. For her dress I'd chosen a green silk with hand-sewn sequins, copper-colored silk accents, and the Mexican coat of arms—an eagle perched on a prickly cactus devouring a rattlesnake—embroidered on the front. In TV and newspaper coverage of the film, I continued to see people wearing my designs, and I felt elated.

Through my experience designing for *Coco*, I was confronted with the realization that as long as I'm undocumented, I will always be faced with limitations. But it also showed me that I don't need a green card to have amazing experiences. Still, I do hope that someday my immigration status will change, and when that happens, I will finally be able to sew my dreams with a permanent stitch.

Elías Roldán was born in Nayarit, México. He came to the US in 1989. He studied fashion design at Rebeca's Fashion Institute and LA Trade Tech. He is a folklórico dancer and clothes designer. His passion is creating beautiful designs, from quinceañera dresses to mariachi outfits to dance costumes, that celebrate the beauty of Mexican culture. He is the proud owner of Elias Designs, Inc.

Razeen Zaman

Insider-Outsider: Unlearning My Legal Education

As I had no memory of the country I was born in, and the only country I knew refused to accept me, I developed a complicated relationship with the law early on. While I did not grow up with any illusion that it existed to protect me, I still harbored idealistic notions of its transformative power. But it wasn't until President Obama announced DACA that I considered going to law school. Before that, it just seemed like a very expensive way to end up jobless afterwards. I didn't have much exposure to the profession since the only lawyers I knew were the scumbag immigration attorney who stole my parents' life savings and threatened to report them to ICE, and a well-known immigration attorney who advised me when I was sixteen to get married fast to a US citizen. Having decided against being a child bride, I vowed I'd be the kind of attorney that my parents and I had needed.

On my first day of law school, I walked into my criminal law class of eighty people and saw a sea of white faces. I could count on one hand the number of students of color. No, this wasn't in the boondocks where you'd come prepared for this sort of thing—it was smack in the middle of NYC. It was like stumbling into an army of *Game of Thrones* White Walkers but we were the sacrificial idiots who actually took an exam to be in their company. Ironically, the first case we read, *R v Dudley and Stephens*, involving two Englishmen becoming cannibals during tumultuous times at sea, was an ominous sign of the three years ahead.

Law school is really weird. None of it will make intuitive sense. You'll learn about the "reasonable person" standard, which is modeled on how a white man would act in any given situation, which won't make sense to you if you're not a white man. Within your first few days, you'll learn about the Christian doctrine of discovery, which justified European colonialism and genocide against Indigenous peoples by claiming that colonizers had authority to seize lands from and vanquish non-Christians. This incredibly racist and immoral doctrine was adopted into law in *Johnson v. McIntosh* when the Supreme Court held in 1823 that the "savage tribes" had no authority to convey their land without the approval of the US government.

As bizarre as this decision is, it's even more bizarre that this decision has *never* been overturned. In fact, 183 years later, the late and great Notorious RBG cited the doctrine of discovery to rationalize the eviction of the Oneida Indian Nation on land they purchased in the late 1990s, after it was robbed from them in an illegal transaction. Cases like this help you understand how the law cements systemic inequalities—it's called stare decisis, Latin for "to stand by things decided." This doctrine requires courts to follow principles set forth in previous cases when deciding a similar case. If that sounds absurd to you, it absolutely is. If we keep deciding new cases on the basis of prior decisions, how can we ever get to new results? But that's the point. So-called consistency is the rationale whites use to maintain the status quo behind a veneer of impartiality.

In this hostile environment, the students of color naturally gravitated toward each other for protection. We fumed in unison when some student (sometimes the professor!) insisted on repeating the N-word three times in one class period, all in the name of staying true to the text of the case. "But the First Amendment!" White kids love to latch on to the First Amendment when you point out that the language they're using inflicts violence on your very identity. There was a student in my immigration law class, an H-1B visa holder, who kept using the words "illegal immigrant" throughout the semester. It was halfway into the semester when I finally asked him to stop using those words because they were dehumanizing, and well, isn't it obvious that language is often a prime vehicle to justify violence? Apparently not. I had just finished making my polite request when White boy 1 turned to scold me about "intellectually stagnating" the classroom conversation. White boy 2 incredulously wanted to know whether I actually believed words could lead to violence. Why yes, I do.

Our graduation speaker, a chief judge of a circuit court, used her speech to deride the student-of-color activists in universities like Yale who had been protesting for their right to exist in spaces that don't brutalize them. She ranted that the student activists were censoring conservatives, molding campuses into "bastions of intolerance" and intellectually stagnant spaces. The most radical of my law school professors left the graduation in protest. I wish I could tell you I was principled enough to do the same, but after enduring the psychological warfare that is law school, I thought I owed it to myself to walk onto that stage and get my diploma.

The three years behind enemy lines had done its damage. It didn't take long to realize I had become more conservative than I ever could have imagined. Before I went to law school, I worked as a campaign organizer for an organization led by undocumented youths where I coordinated civil disobedience led by undocumented people. By the time I graduated, I had adopted the mentality of a risk-averse attorney who couldn't see the forest for the trees. When an undocumented acquaintance consulted me about engaging in a protest where she could get arrested, I discouraged her: "Yes, I understand you want to do this for the movement, but considering you're my client, my responsibility is to you." I was acting as if I weren't undocumented—that I had no accountability to the movement that had kept me afloat so many times. Before DACA, when I kept meeting dead-ends because I couldn't work without papers, it was the courage of undocumented youths shutting down congressional offices, crosswalks, risking arrest and deportation, that was my life raft. It was an act of deep betrayal to act like the people I was helping were clients rather than members of my own community and therefore needed to be treated accordingly with the traditional lawyer-client divide.

My time in law school had taught me that lawyers are gatekeepers to specialized knowledge, which we use to increase our value. I had hoped that when I began practicing law, I'd be able to transmit some of this knowledge to immigrant communities, thereby empowering them. But direct legal services, I learned, were depressingly unsatisfying. I was representing low-income immigrants who had been through hell, but I quickly realized that practicing immigration law requires upholding its racist tenets—an insistence on humiliation and victimization as the primary modes through which you can request forms of relief. In our warped immigration system, the whites slaughtered Indigenous peoples and manufactured their citizenship on the basis of manifest destiny, before creating categories of inadmissibility and deportability. The rules are different for the rest of us. We must beg to have a home, despite every reason we need to stay.

Unlike the presumption of innocence in the criminal legal system, in removal proceedings, you must admit guilt to even ask for many forms of immigration relief. You will end up kneeling before the government in some kind of a hybrid between a very high-stakes college admissions essay and a confessional. Then, you tell the judge some variation of "my shithole country isn't where I want to live so I will tell you the most invasive details of my life and beg to prove my worth before you leave your job at 4 p.m. today." The law, cunning in its façade to remain objective, won't call another country a shithole country outright. But the implication is the same.

The law has a stagnant framework for migrants seeking status—it doesn't matter how complex your life is. You tell the judge the same offensive narrative, rendering yourself a victim awaiting rescue by your savior: America. The more trite the narrative, the stronger the case. There's no room for multifaceted truths in this narrative—no room to acknowledge that you could fear your country but still dream about going back. I represented LGBTQ-identifying individuals from Guyana, Jordan, China, Russia, Kenya, and Bangladesh. The paperwork we filed and the arguments we made for each of these cases were very similar. It's a blueprint that attorneys never discard because we know it's successful: emphasize how hard-working your client is (we call these "equities"), how they escaped a horrible life somewhere, usually in the Global South, and how the United States is the antidote to their suffering.

The problem, however, for an attorney who tells herself she's doing social justice work is that this narrative ignores how the United States' foreign policies, trade agreements, military interventions, carbon emissions, or funding priorities have created conditions in other countries that compel migrants to come here. As a matter of basic fairness, you'd like to tell the judge that your teenage client from El Salvador who narrowly escaped the 18th Street gang should be granted legal status as a way to hold the US accountable for its role in mass deportations of gangs that originated in the US. One of my first clients, a Vietnam War veteran and legal permanent resident, was funneled into the deportation system after he had been racially profiled and caught with drugs. He started taking drugs to cope after he returned from the war with post-traumatic stress disorder. The narrative I wish we had gone with was that he was a victim, not of another country, but of US imperialist ambitions and racism. Instead, we emphasized how apologetic he was about the incident, his rehabilitation, and the fact that he had been a tax-paying, productive member of society for so many years. The reality is that immigration judges have a lot of discretion, and I wasn't willing for my client to risk deportation for an experiment on how to craft a more critical narrative.

In addition to the constant retellings of the same flawed narratives, noncitizens in removal proceedings must perform their victimhood. My clients understand that their grief, manifested into tears, is expected in a courtroom to back up their testimony. I represented an asylum-seeker who fled Burundi after soldiers stabbed and almost killed her when they were searching for her colleague, who had just announced his bid for presidency. We rehearsed multiple times before the hearing, and by the time her court date arrived, her answers to our prepared direct examination sounded rote, and I was concerned that the immigration judge, a former prosecutor from a conservative part of New York, wouldn't believe her. When the judge seemed skeptical, even after seeing photographic evidence of the stab wounds, my client broke down in the courtroom. I was surprised because during the many months of preparation for her hearing, she hadn't cried once. Shamefully, once the tears began to roll down her cheeks, the first thing that crossed my mind was not her well-being but rather that this would be strategic in persuading the judge. The truth is she not only understood the psychology and culture of the courtroom intuitively, as is often the case for those who are most subject to power, but she also knew to preserve her emotional energy for when she needed it most.

You would think that someone who shares the same marginalized identity as her client would be more careful to subvert pernicious narratives and performances. But when you've been subjected to the most racist education possible and then you practice an area of law that relies on your legitimizing white supremacy, you slowly realize that you've become a soldier for it. Every argument you think is helping your individual clients is, in reality, undermining the dignity and power of your community. I once prepared a gang-rape victim from Guatemala for a hearing, and she kept forgetting the order of events of her rape until she finally burst out, "Just tell me the right answer!" She was looking at me as if I were the immigration judge or ICE attorney. I defensively told her that a linear chronology was the only way the judge would understand the chain of events, but the truth is that I had become so entrenched in the flawed ideology of the legal system that I was doing the work for it, traumatizing my own client in the process.

Despite the harmful narratives that I've upheld, I have witnessed the real-life consequences of helping someone obtain legal status. I've helped stop deportations that I know have dramatically improved the trajectory of individual lives. But this does not resolve the dilemma that the sole narrative allowed by the law is one that upholds American exceptionalism at the expense of the Global South. It also does not explain why America demands immigrants render themselves as powerless in order to grant them legal status.

I recognize that the legitimacy I am afforded as a lawyer within the legal system has led me to compartmentalize my identities so that I regularly validate the same system that renders me unlawful. Realizing that I had lost the defiance necessary to be an effective resistor of the system is what led me back to organizing. I now organize with a group called RAISE (Revolutionizing Asian American Immigrant Stories on the East Coast), composed of undocumented Asians. In this space, the lessons I've learned in law school are of no assistance. We don't separate emotions from logic here. Our emotions are what guide us toward goals that seek to reclaim our voices from a system that sees us as powerless lawbreakers.

I wish I could tell you I figured out a way to radically practice immigration law in a way that does justice to the complexity of migrants' lives, but law as it exists makes that impermissible. Until law can be wielded as a tool of morality, written from the lived experiences of the marginalized, it will never free us, and we must look elsewhere for justice.

Razeen Zaman is a South Asian immigrant who grew up in Queens, New York, where she's lived for more than two decades. She works as an immigration attorney during the day and organizes with RAISE. Her favorite things to do include watching *The Great British Bake Off* and then attempting to re-create complex baked goods, which 99 percent of the time don't come out as planned. Although some suggest she stick to lawyering, she will not be dissuaded.

Alexa Vasquez

Querida Zoraida

How are you, mujer? I miss you, *gurl.*

It's been six years since you've passed, mana
And still memories of you smell like hot chamomile tea with honey
left cooling on top of an overcrowded side table
And still memories of us laughing over burnt tamales
makes my eyes water
my throat tightens with purple melancholy

I have taken fewer bus rides down Bristol since you've been gone
But on the days that I do
I press my nose on the muggy window of the Cuarenta y Tres
hoping to spot you walking carelessly with bags of groceries

A gentle giraffe in the middle of Santa Ana carrying
Warm bolillos & leftover Hawaiian macaroni salad

One less passenger on the forty-three
One less Queen performing at the Jack in the Box
  Christmas Party
On the corner of Civic Center

Santa Ana became a completed memoir when you left,
  titled:

Y Todos Me Miran by Zoraida Reyes and Alexa Vasquez
A celebration of sisterhood; woven together by migration
And home in our own bodies

My love and grief brushed on a 34-by-44 canvas for you
Titled Zoraida, La Muerte y El Pajarito
A metaphor for a memory:
    I remember walking down Fairview towards 5th Street
    You suddenly stopped and panicked
    ¡El Pajarito, mana! ¡El Pajarito!
    A baby bird had fallen from its nest and into traffic
    El Pajarito died that day
        I saw fear in your eyes

You visited me in a dream
I'm seated on a bus between two men when it comes to a
  stop
I lean onto the window trying to piece together where I am
There you are
    Wearing your thrifted low-rise jeans and baby tee

Just as young, beautiful, and quirky
As the day we pretended you were a paletera next to an unattended carrito

The doors open as you take your beloved yellow beach cruiser off the rails
I try to run out the bus to let you know I'm here
but you close the doors before I can get off
pouring out a gentle smile you say
aquí no es tu parada, mana
and the bus takes off
Continuing its route

I have now been on hormones for 8 years,
This transition a dangerous swim through the Drake Passage,
A second puberty that has gifted me breasts
Skin as soft as my Oaxaqueña mother's
You and I sculpted from the same clay
By the hands of Aphrodite

We both shared a deep understanding of the hurt that connected us
Fathers who could not hold us tenderly, immigrant sons
Con sueños de color de rosa

Mom brought me to the United States in '91
Leaving behind Oaxaca, mi abuelita Rica and her tejate

Reuniting my sister and me with our alcoholic father
Whose rare I Love Yous were found at the bottom of a Bud
  Light beer can
His kisses belt imprints on my growing body

Without your mother in Michoacán
You had your own painful childhood memories
I remember the morning she called to let you know your
  grandfather had died
You ran to the bathroom because you could not breathe
Tears dripped onto the floor as if to mop clean what he did
  to you
Roots pulled from your heart on English Street

But mana, we are goddesses, birthed from the foam of a
  man's body
Auténticas Intrépidas Buscadoras del Peligro
Muxes en vela

Spirits like ours never die, mana,
The light we carry cannot be buried like shame
Because even *without papers* we are a spectrum of brilliant
  colors
Burning and dancing

It's Christmas time here. And March at the same time
A pandemic has kept us inside most of the year
Ismael and I celebrated our 4th wedding anniversary

In quarantine in our cozy bed with our four cats
Max, Mushu, Sully, and Melon

He is 5-foot-6, an inch taller than me
I don't mind that at all
In his arms I become a pisces swimming in gemini

I met him in a Paramount Studio lot
We were two extras
In our own love story

His hair is dark black and always combed to the left
His eyes are various shades of kindness
Sometimes when I am staring deep enough
I can see my own reflection

He was born in Michoacán, like you
Once first-born daughter
Now second son
First man I have ever loved
deeply

He has shown me there is life after love
more love, mana

Javi married us under a willow arch in our living room
On a cold and rainy day in November
Florence Welch sang Cosmic Love as
I walked across our carpet barefoot

In a white cotton tiered skirt and cream silk lace blouse
Our wedding cake was a two-layered buttercream
frosted cake from Stater Bros. dressed in pretty market flowers
Grocery marble cake tastes the sweetest when accompanied by joyous tears

Family and wedding pictures sit still on our gray Ikea bookshelves
But what's missing in them is you
I imagine you would have worn a cream Tadashi Shoji lace dress
Paper Forget-Me-Nots in your hair
Pushing your way under my veil leading la víbora de la mar

And on nights like this, when I miss you the most
I like to dance barefoot in my living room to that Robyn song
There's this empty space you left behind
Now you're not here with me
I keep digging through our waste of time
But the picture's incomplete

In this dance we both survive

I miss you.

Author's Note: My dear friend Zoraida Reyes, Ale to her closest friends, died on June 11, 2014. Zoraida Reyes was a beloved sister, mana, and trans immigrant activist from Santa Ana, California. I did not attend Zoraida's funeral or burial. If it had been up to her *chosen family*, we would have decided on a more intimate ceremony. The media followed her murder case as the community sought answers for her death. The interviews came and went. *"We always hear statistics. We always hear about transgender statistics and their numbers. Well, today it wasn't a number. It was my sister. It was my best friend."*

Alexa Vasquez is a Pisces, a writer, and an artist. Her visual artworks are inspired by Oaxaca. Her writings are memories of growing up in an immigrant household, leaving home, transitioning, and exploring trans womanhood. Alexa, a Voices of Our Nation (VONA) alum and a 2013 UndocuWriters Fellow, has been featured in *Pariahs: Writing from Outside the Margins* (2016). She graduated from Santa Ana Community College in 2020 with a degree in fashion design. She now lives in Corona, California, with her husband, Ismael, and their four cats.

NEW BEGINNINGS

Ola Osaze

things you remember when an orange-hued fascist with a blond combover is voted out of office

1. white women voting

four years ago, in the houston suburb where you live, you wheel your shopping cart past an elderly white woman examining a can of soup at the local HEB supermarket. while filling the tank of your two-door at a nearby gas station, a white woman at the pump next to yours orders her two rambunctious children in the backseat to calm down. in line at the post office, two women—one brunette, the other blond—chatter loudly about their honor roll high school kids and the menu of colleges that awaits them. you picture these women in the voting booth poised in front of the computer, the presidential candidates' names displayed before them. perhaps the elderly woman lifts up her glasses to get a closer look at the screen. or the mother, now sans her two kids, purses her thin lips and stares blankly for a breath. or the blonde and brunette sweep their hair aside while blinking against the harsh glare of light. their pale hands caress the mouse, direct the black arrow cursor to the name, then click on the check box next to it. and just like that, they legitimize a candidate who promises to unleash an avalanche of deportations, militarized police, prison bars, and towering impenetrable walls. you wonder what part of this answers the question of what makes for a better life.

2. dancing in philly

days after his inauguration, you wear a pink flowery button-up that screams *gay*. together with three of your friends you wind your way down a philly sidewalk crowded with people caught in the full grip of saturday night. tunde, all mesh top and skin-tight pleather pants, gives your procession a soundtrack. *caro your body necessary ah necessary, caro carry leave story. ah leave story. caro dey make my head dey turn.* his fingers snap out a syncopated rhythm. beside him, myra undulates, lithe and free. under the yellow glow of the streetlamps, she is incandescent. you catch eddie's hand and send her spinning. she is an electric blue cloud as her halter swirls around her. then she stops and does a frenzied azonto, to which you all whoop and *ayeee.* you do the shoki in circles even as passersby gawk at your troupe and spectate. in response to the questions and curiosity splashed on their faces, you want to hurl "tonight we dance to shore up our reserves." sometimes in the face of that which is incomprehensible, all you can do is dance.

3. human beings

in nigeria, the parliament grandfathers anti-queer penal codes inherited from british colonizers. in zimbabwe, an old freedom fighter turned dictator calls homosexuals dogs. in uganda, an american pastor rallies the government to hand out death sentences to queers. in america, trump denies entry to nigerians, somalians, sudanese, libyans, iranians, syrians, and yemenis. he lowers the refugee ceiling and seals the border shut. migrant children are shuttled into concentration camps, while haitians are disappeared into dark dank cages. how do you illegalize a human being?

4. clubbing

at the club's entrance, you, eddie, myra, and tunde join a quickly moving line, presenting your african passports with shaky hands to security guards who run their hands over your bodies and shine flashlights into myra's and eddie's purses. just before you enter through the double doors, myra is suddenly nowhere to be found. you see a bouncer scrutinizing her passport, slapping it shut before handing it back to her. she comes around on the other side of the ropes they've used to demarcate the line. "he asked me why my passport had male instead of female, then he said my visa has expired so i can't go in." you all peer at her passport. the visa granting her entry to the u.s. has indeed expired. "are you walking around with this?" tunde asks her, incredulity written into his gaping mouth. myra shrugs, a forced calmness belied by the terror evident in her eyes. "but your passport has not expired so he shouldn't be refusing you entry," quips eddie as she flicks her long braids to the other side of her face. "can we go talk to him?" she asks. "no," myra says through clenched teeth, "i don't want any trouble. i'm fine with going back to the hotel." all four of you are frozen at a crossroads. you could go back to your hotel, sequester in your rooms far from security guards acting as immigration officers. you've grown used to running and hiding.

5. land of democracy

a gay man watches an irascible mob engulf his lover in port harcourt. what remains is a bloody pulp, no longer breathing. scraping together all the money he has, the gay man escapes to the land flowing with milk and honey. in cameroon, a man leaves in the middle of a civil war instigated by britain and france. he travels from africa to south america to the great land of democracy. a congolese woman flees rape and domestic violence with her heart in her mouth. all three prostrate themselves before customs border patrol at the u.s.–méxico border. two out of these three die in detention. can you guess who?

6. myra

in her country of origin, myra is bundled into a van, taken into the bush where she is beaten and tortured. she escapes somehow. she arrives in america where her transness still makes her a pariah. the life expectancy of a trans woman of color in america is thirty-five. in 2016, more than twenty-five trans women were murdered, most of them black. *2017 was the deadliest year for trans people in at least a decade*, proclaims a *mother jones* article. *murders of transgender people surpasses total for last year in just seven months*, declares a 2020 blog post. yet the trump machine works furiously to bar trans people access to health care and employment. while cis men make a sport out of hunting down black trans women, famous authors and pundits use their platforms to deny their womanhood. how do you criminalize a body?

7. worth celebrating

you can't remember the last time you felt unafraid. four years of trump have brought nothing but nightmare-tinged sleep; images of black people devoured in a sea of water cannons, rubber bullets, and riot-geared police have left you staring into the dark of your room at 3 a.m. gasping like a fish. when biden's win is confirmed, you spend an hour on the phone with a good friend, another queer black migrant, talking about the joy you're trying to locate within yourselves. "we've won this battle in the midst of an intractable war," they say in response to your question of "what is worth celebrating in this moment?" together, you celebrate the black women who mobilized georgia and the black people who crammed into voting booths in pennsylvania to turn red swaths of the map blue. you laud the black people everywhere who spilled onto streets in protests months and years earlier, hoisting up *black lives matter* banners and signs. you give yourselves permission to rest in preparation for the next battle.

Ola Osaze is a trans masculine queer of Edo and Yoruba descent, who was born in Port Harcourt, Rivers State, Nigeria, and now resides in Houston, Texas. Ola is the codirector of the Black LGBTQIA+ Migrant Project and has been a community organizer for many years, including working with the Transgender Law Center, the Audre Lorde Project, Uhuru-Wazobia (one of the first LGBTQ groups for African immigrants in the US), Queers for Economic Justice, and the Sylvia Rivera Law Project. Ola is a 2015 Voices of Our Nation (VONA) arts workshop fellow and has writings published in *Apogee*, *Qzine*, *Black Looks*, and the anthologies *Black Futures* and *Queer Africa II*.

Oscar Vazquez

Body of Work

May 2009, Sun Devil Football Stadium—The big day had arrived, the day I'd been waiting for since I started my mechanical engineering studies four years before—graduation. I stood on the infield under the blistering Arizona sun. I eagerly looked up at the reserved sections hoping to see a familiar face. My family hadn't arrived yet. The seats were filling with people carrying congratulations signs, balloons, and vuvuzelas. Alice Cooper was onstage playing "School Is Out" in the background. Looking around, I could sense the jubilation. The graduates were walking around as if floating—the pressure of demanding classes and grueling tests had lifted off their shoulders.

But for some of us, the mood was different. I was suffering from doubt, from fear of the unknown after graduation. Yes, I was graduating with a degree in mechanical engineering from Arizona State University, but I was still an undocumented immigrant from México. I would have a degree, but no way to get a mechanical engineering job in this country.

Something else added to my anxiety. A few feet away was President Barack Obama, our commencement speaker. I was one of the graduating students invited to be onstage with him because I was getting special recognition at the ceremony.

ASU saw me as a symbol for the potential that all students have—particularly those without documents.

When President Obama took the stage, I listened carefully to his words. "And I want to say to you today, graduates, Class of 2009, that despite having achieved a remarkable milestone in your life, despite the fact that you and your families are so rightfully proud, you too cannot rest on your laurels. Not even some of those remarkable young people who were introduced earlier . . . You can't rest. Your own body of work is also yet to come."

While Obama spoke, I thought about my body of work so far. What had I accomplished up to this moment?

December 1998 started off like any other December. Only a few weeks were left of school before Christmas break. The weather was cold, and oranges, pecans, and peanuts were being sold everywhere. The season of Christmas ponches, piñatas, and posadas had arrived. I was twelve and attended middle school in the little town of Temósachic. In the native Raramuri language this means "the place of fog." The town is located in the mountains of the Mexican state of Chihuahua. Everyone I cared about lived here—except for my dad. He lived in Phoenix, Arizona. He'd been gone for about a year. He wouldn't tell us that he missed us, but I could hear the longing for us in his voice whenever he called.

When I came home from school, my mom sat me at the table and told me we were moving to Phoenix to be with my dad. I didn't want to leave. This town had my whole life in it. I was happy here. I knew we were poor, but so was everyone else. It didn't matter how much I complained about leaving. At twelve, my complaints did not carry much weight, and my fear of mom's chancla had more power. So, one freezing December morning we left my hometown. I cried, though to this day, I don't like to admit it.

We made our way to the border town of Agua Prieta, Sonora. We drove to the edge of town on a dirt road. There was a fence separating Agua Prieta from Douglas, Arizona. The border fence was tall and intimidating, but it didn't run forever. A few miles outside of town it shrank into a less imposing barrier. It was there that we found a hole to crawl through. I don't remember much about the next few hours—not because the memories don't exist, but because the paralyzing fear of being a defenseless twelve-year-old in a place I didn't know with two smugglers I didn't trust makes it painful to relive the memories. There is the walk through the cold desert, the Walmart in the distance, and the warmth and relief I finally feel when we walk into the store's garden center. A few hours after that, I am in Phoenix with my mom and dad, a happy family once again.

Within a few days, I started going to an American middle school. I had always liked school, mostly because of my competitiveness—I liked to have the best grades. This school was different, especially the culture, the material we learned, and how we were taught. In México, teachers were revered, viewed with respect. In this middle school, students talked back, did not turn in homework, just daydreamed and ditched classes. In my little hometown, I had one classroom, and teachers would rotate in and out. Here, I had to switch classrooms and would get lost in the hallways. There was also the communication barrier. I had no idea what I was being taught most of the time. I couldn't understand English and would hear only one really long word—like a never-ending train passing me by. How did my teachers talk so fast? Even the history I was learning was different. Growing up I'd been taught that Pancho Villa was a hero of the Mexican Revolution, a freedom fighter, a leader of the people. But now in the US, Pancho Villa was a bandit, a ruffian, and a thief. The best way I can describe this period of my life is "the great confusion." I was no longer sure who I was or where I fit in.

The years went by. I learned English, graduated middle school, and ended up going to Carl Hayden Community High School. It was there that I started to look for places where I would fit in. I ended up choosing elective classes in marine science and the Junior Reserve Officers' Training Corps. I fit in to the JROTC curriculum so well that by my junior year I was the leader of the adventure training team. We would go camping and hiking, we learned how to read maps and how to perform basic first aid, and we even got to visit a few military bases. My JROTC instructors were both Vietnam War veterans. I wanted to walk with my head held high like they did. To be proud of serving my country. Even though I didn't have papers, I felt like I was an American. I wanted to join the military, learn how to jump out of planes, and be a soldier. But I had come to this country through a hole in a fence, so I had to look for something else I could do.

I found robotics in marine science classes when we were tasked with designing and building a robot. My high school would compete with this robot at a national competition in Santa Barbara, California, against powerhouse colleges and universities, including MIT. I didn't know what MIT was, so it didn't intimidate me. I just wanted to work with my classmates and build the best robot we could.

In June 2004, the Marine Advanced Technology Education ROV competition took place. Unlike MIT's robot, which cost more than $10,000 to make, the little robot that we named Stinky was made out of PVC pipes, cheap trolling motors, aquarium pumps, and even milk jugs we found in the recycling bin. Despite being made of cheap recycled or spare parts, Stinky actually worked. It did everything we wanted it to do and then some. My teammates and I—most of us undocumented immigrants—won first place in the competition.

The national championship changed my life. It was because of this competition that I was able to attend Arizona State University and major in mechanical engineering. My robotics teachers, Mr. Lajvardi and Dr. Cameron, were pivotal in my choice to pursue higher education. They were my introduction to engineering and applied science. Through their teaching, I was able to better understand the scientific process, to value math, and to see failures as opportunities to learn. I learned that it is not a good thing when electronics start smoking, but, most important, I learned that challenges could be overcome through teamwork and the application of science.

But as I sat onstage that day in May 2009 at my graduation, I realized that the challenges before me could not be solved with science. No formula in my textbooks could help me.

"Your own body of work is also yet to come," Obama had said.

As much as I wanted to build my body of work, my undocumented status would hinder my progress. Like many students, I had to work and go to school at the same time, but unlike other students, I couldn't get a job through the university and didn't qualify for any student aid programs because of my status. I'd worked construction to put myself through college, broiling under the Arizona sun along with other undocumented workers. I'd been open about my undocumented status with the media and had been featured by the US Senate as one of the students who would benefit from the DREAM (Development, Relief, and Education for Alien Minors) Act, although unfortunately to this day, the DREAM Act still hasn't passed.

To me, President Obama's words meant that even though I had worked hard to get where I was now—it wasn't enough. I couldn't rest. I never had and wouldn't start now. I needed to find a way to get my career started. I needed to prove to myself that I could not only do well in school but also be a successful member of society and pay it forward.

Following graduation, I had a conversation with my wife, who had given birth to our first child not long before. She was a US citizen, and this meant she could file an application to legalize my status. However, the law required me to go back to México and apply from there. I knew that once I crossed the border, I wouldn't be able to come back until I had a green card. For some, it could take weeks. For others, it could take ten years. We chose to take the gamble. It was the only way I had to continue to build my body of work.

So, I went back to México to start the process of getting a green card. People don't realize how arduous this process is—how difficult it is to produce all the requested documents, from every tax form for the previous five years, to financial information, to proof of the relationship with my wife, to the translation of my birth certificate. During this time, I missed my daughter's first steps and her first words. She saw me so seldom that she didn't recognize me when I finally returned. My wife was forced to parent by herself, not knowing if I would ever be able to return. The toll the distance took on my family was immense, but our love endured it.

I ended up spending almost one year in México waiting for a green card. The day I received it and made my way back to the US, I remember walking over a pedestrian bridge to El Paso, Texas, my head held high, no fear of the green-striped vehicles anymore. I had the document I'd been longing for since I was twelve years old.

The first things I did when I got back were to get a driver's license and apply for openings for mechanical engineers. On the way home after a job interview, I saw an army recruiting station. I stopped in to see what it was like. Next thing I knew I was flooded with memories of high school JROTC, of my dream to jump out of a plane.

Then one day, I finally did.

By November 2011, I was assigned to a military base in Anchorage, Alaska, and a month later, I was in Afghanistan and had the honor of fighting for my country.

In January 2016, I was invited to President Obama's final State of the Union Address. As I sat in the viewing gallery in the Senate, I reflected on how his commencement speech seven years earlier had made such a difference in my life. When I first met him, I hadn't been sure where my future was headed. Since graduation, I've lived in three countries and five US states. I have held multiple jobs in the army and in the civilian workforce, and I have been featured in articles and documentaries and a book and movie called *Spare Parts*. Today I am a data analyst for one of the largest railroads in the country. I still use math and science, and I continue to build my body of work.

Oscar Vazquez came to the United States at age twelve, when he moved from México to Phoenix, Arizona. He excelled at Carl Hayden Community High School where he was a member of the underwater robotics team, which won a national championship in 2004 that pitted the team against some of the most prestigious engineering colleges in the country. That opportunity led to a college STEM education, where he earned a BSE in mechanical engineering from Arizona State University in May 2009. With help from Senator Dick Durbin, who spoke from the Senate floor about Oscar's case, he was granted a visa in August 2010. Six months later, Oscar enlisted in the US Army. He now works for Burlington Northern Santa Fe Railway, in Fort Worth, Texas, as a business analyst.

Sól Casique

POEMS

Bull Vision

The dull clippers find my cervical vertebrae and shake me
until I resurface
Like a surucucu snake crackling under its skin
Leaving behind the dust of past selves
Buoyant, tossed in boundless bodies of water

The ocean holds me up to the sun for a glimpse
Of the land that I haven't visited in eighteen years
I'm floating near the center . . . the homeland
La Victoria, the city of independence

It plunges me into the soft void
Between gasps, I try to understand this migrant, transient,
fluid body
I don't know who I am,
or where I am

Somewhere in an uncharted galaxy,
I am stuck in an eclipse

Like Hathor, I bridge the boundaries
Between the cosmos and earth without interruption

Where a bull looks at me above the perpetual magnolia tree
Eyes deep and eternal, anchoring me within myself
The tough petals framing its light brown hide
Its proud horns hold the full moon between them

It herds me through a sandstorm that tries to swallow me
I learn to shapeshift
Just like the bull
Unbounded

I move purposefully tearing down the walls
Shaking off the pestering flies murmuring curses to me
Disrupting my sacred path
You don't belong anywhere and nowhere is home

Sometimes I'm in a crowd with other creatures who are lost
Sometimes I find myself alone in the desert once more

Tastes shift as I now travel guided by my own will and intuition
Kneeling before an altar woven out of
Forgotten memories like the sweet smell of plantains
Or the omnipresent fog that glides down the Andes

I trot to the shore
I peer into the waves, the endless stars

Blinking back at me from the depths
I have become bull like

My hide,
Kaleidoscope of epiphanies
A road map of moments where I was led to darkness
Where others learned to fear my transformations

How my horns reflect moonlight
How my eyes become a telescope into the sun's spirit
An infinite honey
Something untameable

If I, an undocumented person, were to own a spaceship or two or three

One is shaped like a blanket octopus—
Conceived by the gravitational forces of supermassive
black holes
It's controlled by its skulless brain,
Cobalt blue and iridescent,
Tentacles dipped in electricity
Changing shape and decorating its form with striking
fauna, dead suns of old
And bright psychedelic clouds that water its simmering
body

Pools of sweat, that splatter onto asteroids, are forged by my
salt and rage
From being continents away from my family
While my abuelo lies alone in a crowded hospital room
A virus feeding on his body
Nightly rosaries and novenas forming a limp shield over his
weak lungs
The world outside roars as his spirit flows over the burning
streets
A nation suffocating at the hands of soldiers, and corrupt
leaders

But the sun won't set today or tomorrow
Or for another five billion years
And so, I make myself useful in this octopus
I call it Deathstalker, after my scorpio rising,
Crackling and hissing between stars
As outlaws of space we adapt to its galactic surroundings
Speeding away   secretive and undetected

This spaceship's body a powerful whip
With no expiration date on its livelihood
It harpoons the belly of enemy spaceships
Whose vessels driven by greed
Absorb and extract the parts of fallen ships to a point of
  no return
We liberate those long thought to be lost
Turning them into tourmaline

Then, there is a sequoia floating within supernovas
An infinite and proud 87-meter giant spaceship
Nebulas are caught in its shallow roots and long branches
I name it Taurus,
Like many trees that have followed me in my dreams,
Its furrowed saffron colored bark is whittled
And carved with the archives of moons
From galaxies that it helped raise and others it watched
  crumble

This sequoia exists carefree in space
Not for the consumption of humans or their plans that only end with downfall
Its existence, for once, is just for itself
Energy powered by pleasure and solar
And I ride it to access the songs of my childhood that vibrate throughout its spine

And I get to remember
Boleros dancing throughout my abuelo's apartment,
Summer air smeared with laughter
My family unaware that this will be the last time
We dance on the beige and blue flower tiled floor

I get to daydream   time travel   become ancestral anthologies
Guardian of wanderers and rainbow children
I get to be infinite

Amorcito, the dripping skull, is the last one
Two heart pickaxes cross on its crown
Drenched in various tongues of inferno orange,
I sail it through Jupiter's hurricane and Saturn's rings
I coat Mars's red sand across my chest for protection
Flying freely and borderless
For there are no checkpoints, or ICE agents asking for IDs or papers
There are no prisons here
Only a birthplace for golden roses

This ship has a lingering incense of nostalgia, of sacred
moments
Sustaining this mortal body—
6 a.m. aroma of freshly made arepitas
Sprinkled with homemade queso blanco and caraotas,
Immaculate white-bowed shoes running through the
plaza,
Tiny strawberry ice cream cone that once melted down
my small arm,
The melancholy notes of the llaneras hitch onto the
breeze that carries these smells,
Taking it home to the Venezuelan plains

I'm a space rider now
An alien without chains, dressed in talismans and fire
Observing the depths of my homeland
Wondering what price I paid for this freedom

I tighten my leather jacket
A proud bull embroidered on the back
I put the ship into drive
A hummingbird skull on the dashboard clinks against the
glass
Drifting across the Milky Way
Someone beckons me to *come home*
One sharp right turn and I could be *there*

Surrounded by warm soil that my worn feet know too well
I taste the pabellón one more time
I embrace my abuelo one last time
Because nothing is permanent and everything is deserving
and divine

Sól Casique is an undocumented creative, centering their work on themes of transformation and the beauty of being, creating, and imagining beyond the limits of colonialism. As a stellium Taurus, they're constantly craving birria tacos, a warm window seat to read by, a sweaty perreo mix, and the fall of white supremacist states. There's no negotiation for their identity.

Barbara Andrea Sostaita

Undocumented Success Story

You were six years old when your family arrived in the United States—confused, curious, convinced your parents brought you here to meet Mickey Mouse. Your flight landed in Miami, and immediately you began searching everywhere for the wide-eyed mouse and his unmistakable yellow shoes. But he was nowhere to be found. Your parents led you to believe you were here on vacation. But the months passed, and your tourist visa expired, and you kept asking when you'd go home to Argentina. "Pronto," they said, over and over again. They never confessed that you were here to stay or that you were here sin papeles. Not explicitly anyway.

You never wanted to live here permanently. You hated riding the big yellow bus and preferred walking to school hand in hand with your dad, stopping at the bodega to charm the salesclerk into giving you a pastry, free of charge. Everything was so quiet here, so still. Not like Buenos Aires, always abuzz with the sound of your primas's laughter and your abuelita screaming when she couldn't find the paprika for her empanadas and your dad bringing you along to demonstrations, chanting "Viva Perón" down the cobblestone streets. "Buenos Aires breeds revolutionaries," he told you during one rally.

What he never told you was that many of the revolutionaries had to flee Argentina, and that many of the ones who stayed have been disappeared by the state. His own family left in the 1980s, during a military dictatorship that turned the ocean into a graveyard and the streets of Buenos Aires into a parade of mourners. He also never told you how much the dictatorship still haunts him. He will not confess that it's part of the reason you left Argentina, that even now—twenty-two years later, when you're no longer undocumented—he still refuses to return to his country. You call it his country because you were never given the opportunity to call it yours. The Argentina you remember is only a faint trace, a ghostly haunting. There is little to miss because there is little to hold on to. Your primas grew up without you, and your tía passed away without you, and the country you left has changed irrevocably.

You've changed, too. You no longer mind the stillness. You've learned to call the United States home. At age twenty-eight, you have a doctorate in religious studies from one of the best schools in the country. You are an undocumented success story. Or so your parents tell you. And their friends. And their congregation. And your high school teachers. And childhood guidance counselors. And the well-meaning folks you met through your dissertation research. And fellow migrant youth. And your neighbors. And. And. "¡Sí, se pudo!" they comment on your Facebook status announcing that you defended your dissertation. But those words—meant to be celebratory, to suggest overcoming obstacles—strike you as static and predetermined. It erases the constant self-doubt, the ongoing anxiety, the endless pressure to prove your worth. It's not sí, se pudo. It's a never-ending struggle of whether or not si puedes seguir, cuánto puedes aguantar. You consider replying to the comments and confessing that success almost destroyed you.

In 2010, you enrolled at a women's college only miles from your parents' house. It was the only college that would accept you, and you could live with your parents to save money on campus housing. You missed out on the traditional college experience, but you didn't mind. After years of panicking about college and grieving your dreams of pursuing a higher education, you simply felt lucky to be sitting in a college classroom. After all, you were taught that higher education would save you and prove your worth to this country. Nothing else mattered. With a college degree, you would show that immigrants belong in this country. You would honor your parents' sacrifices.

Four years later, you moved six hundred miles away for graduate school. Your dad sobbed as he drove away in an empty U-Haul truck after helping you move into a one-bedroom apartment in New Haven, Connecticut. That's the first time you remember seeing him shed tears. He confesses that the day you left Buenos Aires, during a layover in Chile, he briefly abandoned your mom in the airport food court so that he could cry in a bathroom stall. Part of him understood that he would never return to his homeland. Perhaps leaving you in Connecticut reminded him of that hour in Chile, and he realized that this migration was also a kind of transformation and that your relationship would never be the same again. Migration is an endless string of losses and departures. Before driving away, he told you to remember the struggle and sacrifice it took to get here. That everything he's done has always been for you, which every child of immigrants knows is code for a debt that you never asked for and will always remain unpayable. He reminded you that other young undocumented kids look to you as an example. And all of these words were an immense burden to carry.

You never felt more alone than the years you lived in New Haven. The seemingly endless winters cast a pale gray hue over your days, and like the novels you read in high school, the landscape's barrenness appeared like a metaphor for death and loss and yearning. The days were short and the course load intimidating. You survived weeks on end eating frijoles and rice. You struggled to keep up during seminar discussions, unfamiliar with words like "dialectic" and "epistemology." You felt lost reading theory and you felt like a failure. You sobbed in a seminar classroom after a session of "Religion and the New Spirit of Capitalism," during which a classmate deemed Argentina a failure and called the country's economy "hopeless." You began to distrust that an education would save you. But you couldn't share that with your dad, or with most people, because undocumented success stories are supposed to be grateful. You are taught to count your blessings and pull yourselves up by your bootstraps and be an inspiration to others. You are not supposed to fail, to express doubt, to disappoint your communities. You remember that your parents gave up vacations and drove broken-down vehicles and wore their bodies down working minimum-wage jobs so that you could enter the halls of academia.

You wish the people who comment "Sí, se pudo" on your statuses understood what no se pudo along your journey in higher education. What you had to give up to be their poster child. What you surrendered to succeed. Besides the parties you avoided because of a fear of policing and deportation. Besides the study abroad trips you were unable to join because of your legal status. Besides the friendships and relationships you neglected because time was reserved for your education. You wish you could express how holding yourself to impossible standards and worrying about disappointing your community has damaged your mental health. You wish you could record your conversations with your therapist, the discussion of panic attacks and imposter syndrome. You wish you could tell them that your partner has to frequently jolt you awake at night because the sound of your gnashing teeth frightens him. You wish you could tell them, to paraphrase Sandra Cisneros, that at the end of the day, books can't hold you or offer you a healing touch.

You wish you could tell them how, days after receiving your PhD acceptance letter and your family celebrated you with carne asada, your pregnancy test came back positive, and that immediately, you knew an abortion was the only option because an unexpected pregnancy out of wedlock was not an undocumented success story. Unlike a diploma, a baby would not make your family or community proud. It would be nearly impossible to begin a graduate program with a newborn, to nurse an infant all night and attend an early morning seminar. You do not regret your decision to have an abortion, and you're grateful for the health care you received. But sometimes you dream of an alternate journey, one where you deviated from the plan or you allowed yourself to imagine otherwise. To this day, you wonder what your child would look like and what their first words would have been and the lullabies you would have sung to them before bed. Not much is left of Argentina in your memories, yet the songs you inherited from your grandmother are an exception. You wonder if choosing to keep the child would have allowed you to write a different story, one that refused respectability and success and achievement. Or that defined those terms differently. You wonder who you might be without the pressure to prove your humanity to a society that wanted you disappeared and deported. You fear committing these words to a page, because your parents still don't know about the pregnancy and learning this truth might undo the undocumented success story they've memorized and narrated for the majority of your life.

But you never wanted to be an undocumented success story. You were taught that success would save you, that a diploma would shield you from deportation. And you hope for a world in which undocumented kids are allowed to be human, a world in which you could experiment and fail without the burden of being someone else's dream. You imagine that your relationship with your dad would be healthier in this world, that you would be quicker to forgive yourself when you make mistakes, that he wouldn't put so much pressure on you to be perfect. Perhaps you wouldn't resent him for turning you into a hero, and you would understand that he survived an impossible situation only to come out the other side broken. You realize that unlearning the undocumented success story requires that you heal from your wounds, and maybe that might help heal him, too.

Barbara Andrea Sostaita is from Argentina and holds a PhD in religious studies from the University of North Carolina. She is currently working on a manuscript based on her dissertation, an (auto)ethnographic experiment titled *Sanctuary Everywhere: Fugitive Care on the Migrant Trail.* Focused on the Sonora-Arizona borderlands, the book documents moments of care and intimacy that expose the impermanence and instability of border militarization. Her writings on sanctuary and migration have appeared in *Bitch*, *Teen Vogue*, and *Remezcla*, among others.

Reyna Grande

Not So Sweet Valley

As I was brushing my ten-year-old daughter's hair, she turned to me and said, "Mami, when are we going to Hawaii?"

"Why do you want to go to Hawaii?" I asked.

"Because all my friends have been there, and I haven't," she replied.

I remembered when I was her age asking my father a similar question: *"When are we going to Disneyland?"* My father had laughed and walked away. He was a maintenance worker at a convalescent hospital in Los Angeles, supporting three kids on minimum wage.

Thankfully, at this point in my life, neither Disneyland nor Hawaii is out of my reach.

What is beyond my reach, though—at times—is coming to terms with my children's childhoods being vastly different from mine and accepting that the distance between me and my children was of my own making.

When I was in junior high school, the librarian at my public library handed me books from the young adult section to take home. I walked out of the library with a stack of titles such as *Sweet Valley High*. The girls on the cover, with their blond hair and blue-green eyes, stared at me and smiled, as if they wanted me to be their friend.

Transfixed, reading under the covers with a flashlight, I entered the lives of identical twins named Jessica and Elizabeth. Two beautiful, slender girls with a California tan and dimples on their left cheeks. Their father was a lawyer; their mother, an interior designer. The girls were high school cheerleaders with lots of friends and cute boyfriends. I was an undocumented Mexican immigrant. Four years earlier, my siblings and I had run across the border to begin a new life in Los Angeles. We lived with my father and my stepmother in a one-bedroom apartment. We slept in the living room, my sister and me on the sofa-bed, my brother on the floor.

When I discovered *Sweet Valley High*, I had just completed the ESL program at my junior high and was now in regular English classes. Though my reading and writing skills were good enough to be in a classroom with native English speakers, my pronunciation lagged far behind. At school, my peers laughed when I spoke in class, and so I preferred reading. When I read, no one but me could hear my "wetback" accent. And at least the Wakefield twins never laughed at me. I knew in real life I could never be their friend—there was no one like me in their circles—but my library card gave me the privilege of at least peeking into the lives of these all-American girls.

But I wanted them to see me, too. I wanted to feel that I could belong in their world, instead of being on the outside looking in. I wanted to know what it was like to have two successful parents, to go on trips, to have beautiful clothes to wear. To never want for anything. These books gave me access to something I didn't have access to in real life: white, middle-class America. But as much as there was pleasure in reading about an American life that could never be mine, there was also pain. Lots of it. *This is what I do not have. This is who I am not. This is who I can never be. This is what I will never do.*

I worked hard to be able to have that life one day—but most important, to give those experiences to my future children. Even before I had my son and daughter, I was striving to make sure I could give them the childhood I had never had. My first act of love as a mother was to remove from my children's lives the labels I grew up with—low-income, immigrant, English language learner, first-generation college student. My son and daughter will never face the daily struggles I encountered as I fought for my place in this society, for my right to remain and become a part of the fabric of this country. But neither my college degree, nor my writing career, nor my perfect English prepared me for the experience of raising two American-born, upper middle-class children.

I was born in the second poorest state in México on a dirt floor, in a shack of sticks and cardboard. My children were born in a private hospital in Los Angeles, California. I spent the early part of my childhood separated from my parents when they immigrated to the US without me. My children live in a stable home with two loving parents who have university degrees and professional careers. The first time I ever traveled out of my hometown, it was to Tijuana—to risk my life running across the US-México border. My children have traveled to places all over the US and abroad. Unlike me, they are growing up with "when you go to college," not "if you go to college." They have college trust funds that my husband and I contribute to each month. When my daughter used to play with her Barbies, her dolls were in college or in a book club.

I don't know how to reconcile my poverty-stricken childhood and my children's childhood of abundance. My success in this country has allowed me to give them the life I'd once dreamed of, but in my quest to spare them the trauma of growing up poor and on the margins of American society, I overcompensate and overindulge.

The feelings I had when reading *Sweet Valley High* are the same ones I have now when I look at my kids—the feeling of being on the outside looking in.

One day, I came home to find my husband and kids sitting on the living room floor surrounded by piles of clothes. "What's going on?" I asked as I walked in.

"We're decluttering," my husband answered. He'd recently read Mary Kondo's book *Spark Joy: An Illustrated Master Class on the Art of Organizing and Tidying Up*. He pointed to two piles and said those were the clothes the kids were keeping. "And that one is for donations," he said, pointing to the biggest pile. I gasped when I saw the shirts that I'd bought my teenage son on my recent trips, dresses that I'd carefully picked out for my preteen daughter. They were almost new and still fit perfectly. When I decluttered, I removed from their closets only clothes that they had outgrown or were too stained.

"There's nothing *wrong* with these," I said to my daughter, picking up the dresses and rescuing them from the pile. "And they still fit you. Why would you get rid of them?"

"I know, Mami, but they don't give me joy," she answered.

I flashed back to my Mexican childhood and pictured myself as a little girl walking on the dirt roads of my neighborhood, barefoot and semi-naked or dressed in rags. The few dresses I owned were stained, worn, and peppered with holes. In a moment of desperation, while having to take care of us, my grandmother had even made my sister and me dresses out of a tablecloth, saying that her table had no use for such things.

I was angry, yet, wasn't I the one who'd bought all those dresses? Too many trips to the store to count. I thought of the pleasure I'd felt as I looked through the clothing racks, imagining myself as a little girl wearing those dresses. How could I be angry at my children for discarding things they didn't ask for in the first place? My anger was replaced with guilt—and then, shame. My children are living the middle-class, privileged American life I'd dreamed of giving them, aren't they?

We didn't go to Hawaii as my daughter wanted to. Instead, we went to Europe. I was invited to give a reading of my books in Germany at the University of Münster, and I took my family along. My husband, my children, and I stopped in Paris for a few days before making our way to Germany. After Münster we went to London, Edinburgh, and Reykjavik. My children had traveled to many states in the US, to México to visit my relatives, but this was their first trip to Europe. In Paris we took a river cruise on the Seine and visited the Eiffel Tower and Notre Dame, we watched the musical *Wicked* in London's West End, we had breakfast at the Elephant House in Edinburgh where J. K. Rowling worked on some of her

Harry Potter books. These moments were more surreal to me than they were to my children. *I've brought my kids to Europe*, I kept thinking. Then, I'd think back to my childhood, how when I was a young girl my father couldn't even afford to take us to the movie theater down the street.

On the second day after we'd arrived in Europe, my teenage son began to say, "I want to go home." This became his mantra for the two weeks we were traveling. My daughter didn't complain during the trip, but afterwards, when I was reminiscing about the trip with her and asked whether she would want to go back to Europe someday, she said, "Yes, but not to Paris. I prefer the French countryside."

Sometimes, the things that come out of my children's mouths make me think that I should reread *Sweet Valley High* so that I can understand them better.

There are moments when I feel that my children have gone to a place to which I cannot follow. It is only now that I understand how my father felt watching his children do the same.

When he snuck us across the border to live in Los Angeles with him, he risked our lives because he wanted to give us a shot at a better life by taking us to a country where that was possible. Little by little, my siblings and I became seduced by American life. Once we learned English, that was the language we spoke with one another. Instead of telenovelas, we watched *Small Wonder*, *Beverly Hills Teens*, and later, *Beverly Hills, 90210*. I joined my school's marching band and played the alto sax, while my sister did modern dance and went by "Maggie" instead of "Magloria."

"My children don't speak to me. They've learned another language and forgotten Spanish!" While my father sang along to songs by Los Tigres del Norte and wallowed in his nostalgia for México, my sister and I bleached our black arm hair so our brown skin could look a shade lighter. "They think like Americans. Deny they are Mexicans even though they have my skin color."

The song "La Jaula de Oro" (The Golden Cage) captured my father's reality as an immigrant parent—the price that he paid for his dream to give his children a better life. Did he ever regret bringing us to this country? There were times when he did. But if he could choose between having us live in the stark poverty of our hometown or live in the US, though it meant watching us go to a place where he could not follow—succumbing to assimilation—I know he would always choose the latter.

Whenever I go into panic mode at seeing my children becoming *Sweet Valley High* characters, I whisk them away to México. I take them to Iguala, my poverty-stricken city. We stay at my aunt's house where there is no running water, and my kids must dump a bucket of water into the toilet to flush it or heat up a pot of water on the stove if they want a warm bath. On one of our trips to Guerrero, I carried my camera on a morning walk around my old neighborhood. As I walked past a shack made of sticks and cardboard like the one where I had come into the world, a barefoot little girl jumped up on the gate and said, "Hola!" I returned her greeting and took her picture.

When I returned from my walk, I showed my daughter the pictures I'd taken, including the one of the little girl. She looked at it intently and then said, "She looks like me." I looked at the photograph and realized that my daughter was right—with her hair burning brown in the sunlight, the little girl looked very much like her.

"I could've been her," my daughter said.

In that moment, my daughter's words were a soothing salve on the shame of what my American Dream has cost me. I began to realize that I should stop begrudging my children the life I've given them. If the worst that has happened to them is to live a *Sweet Valley High* kind of life, I am willing to pay the price, like my father once did. Even if the story I've written for them feels foreign to me.

Rommy Torrico

Come Back (2020)

REPUBLICA DE CHILE

Rommy Torrico

Rommy Torrico is a formerly undocumented, queer, trans/nonbinary visual artist born in Iquique, Chile; raised in Naples, Florida; and currently based out of New York City. They have been involved in the (im)migrant rights struggle for several years and infuse much of their work with personal experience and the stories their community shares. Over the years, Torrico's work has been included in several publications and exhibited in galleries and museums throughout the Americas and internationally.

César Miguel Rivera Vega Magallón

Return to the Invented Country: A Theory of Return Migration

La casa ya es otra casa, el árbol ya no es aquel /
Han volteao hasta el recuerdo, entonces, ¿a qué volver?

—"A qué volver," Eduardo Falú

When you finally get tired of grasping for success and a dream you have seen spoken about only on television, you will buy a plane ticket during the off-season and plan your Return home. The mental suitcase you have kept by the side of the door your entire life will become a real suitcase. The morning of the flight will consume you with a horrifying euphoria, and the hours you spend on the airport shuttle on the 405 will feel like seconds. The foul urinelike smell of the Budweiser brewery along the freeway will fade into the salt air of the ocean and the jet fumes of the airport. You'll walk through security, place your luggage on the belt, and when you emerge on the other side, in the bowels of LAX for the first time, you'll realize after all these years that this place is just another shopping mall.

There'll be no pang of regret when the wheels peel from the tarmac, no fear when the plane rattles in the currents, as if California is trying to drag you back to the ground. You'll see the Palos Verdes Peninsula reach out to you from the window, begging you to change your mind with its outstretched, flat palm. By the time it is swallowed up by the Pacific, they'll have brought you a Coke, and you will have forgotten about the sunsets at Portuguese Point.

Then, it'll happen, the moments the clouds part. You'll see Home from the window: gray, ganglion cyst with its sinews crawling through every valley and ravine. You won't want to blink; you'll want to brand this moment into your memory forever, the day you think you have finally seen yourself outside Parmigianino's mirror. The world will return to an ordinary logic, up will cease to be down, but you'll look at your wristwatch and see the portent; two hours behind, still living in the past.

From my first step into the United States at four years old, waterlogged from the drain ditches that keep the Tijuana River from overflowing its historical banks, I knew I would Return home one day. Caught between October gales in the Tijuana desert, I wanted nothing more than to go back to Guadalajara and the perpetual springtime of the Valley of Atemajac.

My time in the United States was supposed to be a simple visit, a momentary reunion with my father, whom I hadn't seen since I was born. The thought of Return obsessed me, chased me throughout my childhood as I attempted to assimilate, to live quietly and obsequiously. The inevitability of Return, whether through a knock at midnight in the form of an ICE officer who would take me away, a raid at my father's job, a broken taillight, or an otherwise anticlimactic end to my American flâne meant that no matter how much time I spent in California, it would always be a transitory experience. Everything about California for me was transient: relationships, places, and achievements. The physical residues of those things, the diplomas, the awards, the photos, my car, an apartment, all had the texture of dried flowers in a journal that one day would slip out of the pages and turn to yellowed powder.

What remains, what survives as memories of places and people, exists out of order in an illogical time placed somehow always concurrently with the moment of finally leaving. In Migrant's Time, the present can't exist; everything must be understood as happening always in the past or menacing from the future—that is, always at the moment of Return. When my husband and I stood beneath a brass arch of artificial flowers at the Van Nuys courthouse on our wedding day, I felt tidal locked to México and, inexorably, to the day I would be in that other country. It happened throughout my life; during graduations, the birth of my siblings, the divorce of my parents—the present was uninhabitable.

Anzaldúa was woefully mistaken—the border isn't a wound, it's a vacuum in the center of the heart. In its great emptiness, every sentiment is collapsed, every small feeling implodes until it is impossible to determine what, if anything, has value, reason, or truth.

Illegalized migrants in the United States are in a state of constant mourning for the lives they could have lived and the people they could have been, the families they could have had if only the circumstances had been different. The civil wars, the economic sackings, the coups, the roundups and dirty wars don't stop claiming lives even when we sign a temporary truce, our green card, or our naturalization certificate. The conditions of our original migration pursue us, stalk us like pheasants in the high grasses of our unutterable desires to assimilate and finally vanish from the gaze of politics and the law.

When we are at our most comfortable, when it seems we have forgotten the stench of the drain ditches or the fumes from the tire in the car trunk we once rode in—a voice comes through the wire to tell us our childhood house has collapsed, our hometown has flooded, a loved one whose lips we can still feel against our infant cheeks has died. Suddenly, time closes in on itself again and we are consumed with the hybrid nature of our Return—impossible, inevitable.

As undocumented youth become undocumented adults and outgrow their butterfly wings and caps and gowns, the reality of aging and of the inability to accumulate even the most threadbare security becomes overwhelming. Despite contributing to the fund, there will be no social security to draw from after our prime working years, there will be no title deed with our names on it. For those who have won a cruel lottery and have work permits through DACA, life is experienced in a series of two- then one-year intervals with the threat of life being upended by a failed five-hundred-dollar process always threatening to fall from above. For older migrants, the experience is not much kinder. The dream of accumulating enough money, or sending enough money home to finally retire in their countries of birth, becomes more and more distant as their prime years slip by and the dream of a new roof or their mother's knee surgery vanishes with every rent increase and urgent care bill. Grandmothers, fathers, siblings, children begin to die. Their funerals become personal productions of *Antigone*; the choice between fulfilling our last obligation in mourning, possibly never returning, or remaining in the country we've lived in hurts somehow even more than our loss.

When my great-grandmother died, a MoneyGram receipt in my hand had to stand for the life of a woman who raised me as a mother. A remittance became the synecdoche for all the love she gave to me, which I now returned in pesos. I held the receipt imagining I was holding her hand. I saw my mother, a steel-girdled woman who had performed miracles to make our life in America possible, fall apart for the first time. Time and distance caught us flat-footed. The fifteen hundred miles, two time zones, and twenty years apart made themselves undeniably present.

There, Return found me again, and the act of leaving the country I was told was a land of opportunity became a persistent, intrusive thought. Even if I did not then wish to, stories of ICE raids on workplaces, of survivors of domestic violence detained at courthouses, of friends in the movement seemingly targeted for their activism compelled me to at least consider the possibility before I felt I would be forced to sign my voluntary repatriation.

I knew I could not survive detention; my diagnosed mental illness makes me particularly ill-suited to confinement against my will. The act of Return was an act of desperation to save my own life from the seeping madness and stress of life in the United States after the 2016 election. The fear, however irrational, of detention and deportation began to eat at my sanity until I was moved to action, much like my parents had been in the currency crises and market openings of the late 1980s and early 1990s that suddenly had robbed them of their futures in Guadalajara. The moment I purchased the plane ticket, I felt much of that crisis was over. I would Return, and I would be free of the uncertainty that had stolen twenty-five years of my life.

Miguel Hidalgo Airport in Guadalajara is a portal; the Arrivals gate echoes with the half-century of lamentations from those torn apart by the sudden need to flee waves of violence in its many forms. The customs agent who checks your bag sees your passport was issued at a consulate and refuses to make eye contact. The place of issue, Los Angeles, marks you as yet another of hundreds of thousands of souls who've passed through, bringing not much more than a bag and sometimes unaccented English. The airport spits you out, and you see Guadalajara before your eyes for the first time. You realize that time has not stood by and waited for you to make up your mind and Return. The city has spread beyond the limits of the place you remember and loved. The cane fields and milpas with the ears of corn turned toward the sky are now rows of concrete apartments. The number of shopping malls you pass is comforting if overwhelming, but a Costco parking lot feels like a sobering slap in the face. As the taxi draws nearer to the house where you took your first steps, you desperately seek to recognize something, anything of the place you left. The cobblestone streets have been upended and paved over; the metal shutters of the corner store are scorched (which you will eventually find out was for the owner's refusal to pay an extortion). You think, *Well, at least the stray dogs are still here*, until you realize even their constituent breeds have changed.

But your grandmother will open the door. An uncanny feeling of peace, of belonging, takes over as you drag your bags across the flagstones you walked upon as a child. The sensation that the love, at the very least, has not changed consoles you, comforts you. It envelops you as your grandmother wraps her arms around you, remarking how thin you've become and how tall you've grown. She reassures you that the decision you made was the correct one without saying so, just with a touch on the shoulder.

Then comes the first flash of doubt, the signs that twenty-five years have claimed your grandparents' energies as they fall asleep, slumped over at the table. You'll realize you have never seen your grandmother sleep before that moment; you'll remember she always rose before you did and went to bed after you had already fallen asleep. You'll remember how strong she was. Your grandfather's cane will fall, and the echo will make you shudder. The grandparents you longed to see, they're no longer there. In that moment, you'll understand you have not Returned yet. Because Return has never been about coming back to a place; it was always about Returning to a time outside the Migrant's Time.

Our first migration is just that, a first. The act of being displaced from your home sets you on a path that forever changes you. The desire to Return is the desire to obtain the status quo ante, an impossibility. Return is a fantasy of other worlds. It's the action outside any governing logic to remedy the tragedy of our first exile. It's an attempt to impose choice and agency where our first migration was simply an attempt to survive. Return is the potential to escape being illegal in the metaphysical. If we can't escape the aftermath of the political and economic forces, we can reshape them in our own lives and claim a stake. Return is the insistence on healing. It is one we are ill-prepared to respond to as our energies as a series of transnational movements across continents have rarely contemplated migrants' wants, needs to Return. We often conceive it as a total surrender, a failure, and returnees carry that stigma when they eventually arrive "home." If we believe in empowering our communities, we must contemplate this act of Return not as that stigmatized surrender, but as a stratagem that allows us to survive and prepare for another day in the struggle.

The choice to Return felt like the first choice I have ever been able to make in my life. Everything else before that point was simply a consequence of or a reaction to the choice my parents had imposed on me that macroeconomic policy and five hundred years of colonial history had made for them. Return was the opportunity to correct course, to rescue the meaning of my life from legislative cycles and election campaigns, to be a person and not a political issue.

Return was coming back, not to México, but to myself.

César Miguel Rivera Vega Magallón is a queer, formerly undocumented, Mexican poet and an advocate for migrant, refugee, and returned/deportee rights. Originally from Huentitán el Alto, Guadalajara, Jalisco, México, they resided in northern Los Angeles County for twenty-five years before self-deporting to Guadalajara in 2018. César Miguel facilitated the Antelope Valley Writers Association before becoming a full-time organizer in the immigrant rights movement and pursuing a major in art history at the University of California, Los Angeles. Their short story "We, Saracens," won the first ever Things I'll Never Say prize for fiction in 2013. Their essays have been published in the Huffington Post, Motherlands Zine, Color Bloq, and on other platforms. César Miguel is the current 2021 México Advocacy Fellow at the Rhizome Center for Migrants in the city of Guadalajara.

Dujie Tahat

POEMS

THE FENCE

Throw your arms and hips
over with your legs
all akimbo, a limbo
if you lower the bar
is a rod—I mean really lower it
doesn't matter if you out
on bond the word is
you not from round
and round and round about way; we
the people, we
don't believe you
are my sunshine, my only
my only, my holy moly
rejects bestowed a trust
the rhyme in history returns to
our sound earth
bound feet first step
on the throat of a new nation under
the weather matters less than
the nipping at your heels.

I TAKE MY KIDS TO SCHOOL AND THE PRESIDENT IS SET TO BE ACQUITTED

The drive time from Tukwila to Everett clocks
in at fifty-six minutes. Some light rain.
School starts at 7:45 a.m. I hear Budweiser
is a prolific Super Bowl advertiser.
Remember the frogs.
Let Noxema cream your face. Farrah
Faucet. Joe Namath. [Motion
to subpoena.] Who ever has the votes
will decide the facts. [Witnesses

walk away.] I built
something beautiful and already it's on fire. [Sorry

wrong sound bite.]

The Nasdaq is down
nearly one percent. [Procedural motion.]
Overnight oats. Orange-petalled rain coats.
One child finds the other's monkey
under the bed.

[The polls say.]

Britain is finally leaving.
By early afternoon the whole metro area will be under a
flood watch.

191 Americans have been quarantined
by the CDC. Someone I know lost their child
so their whole family is unreachable.
Someone I don't know lost their child and I can't stop
watching them grieve.

We reflect
on the anniversary of the Muslim travel ban.

["When you come from the Middle East that can be difficult.
We moved up here for a better life, a peaceful life,
but when you come here, you don't leave your
past.
Your past comes with you. Your land comes with
you."]

I swear I don't
need minute-by-minute updates on another parent's grief.

The nurse's strike
is over. [On the side of the public.]

The dog must still be walked
twice a day. [Cloture.] Patients over profit—
yes, but first peel a scab

off the sidewalk. [Per reports.] Kobe Bryant. KOBE!
Gee—two Bryants. Kobe Bryant. Kobe Bryant. Kobe
Bryant.

Six people
are waiting to hear if they have the Coronavirus. [We await

in suspense.] The Mayor has made a deal with the King
County executive who's cut a deal with the union
boss who's sold a wink wink deal to the President
of the Chamber of Commerce who didn't
need at all
to make a deal with the billionaires
who hired
the people who hired the people
who hired her.

The nurses don't represent all healthcare
workers. [Grease in the wheels.] The people

of the United States decided who
their president should be. I mean it's bad but not [________
________] bad. It's worse.

These borrowers
were, unbeknownst to them, in the wrong program.

The Mexican President will now take control
of legal and extra-legal migration to America.

Dujie Tahat

The Way As Promised Has Mile Markers To Guide Us

Pops bought a '78 Pontiac,
a firebird-stamped gold bar
on wheels, spontaneously,
after a conversation with
an aunt's friend—so it went.
In this country, you can
start a trip from any city
heading in the direction
of your destination, certain,
reasonably, of arriving whole
and needing to piss. A pit
stop in Olive, California, is a
good joke the way my dad
put it. Hold it long enough
and a song takes the mind
off messing the leather up.
Sunflower seeds in a Big Gulp
cup. The most beautiful road
in America. One sweeping
hand steers. Sun-drenched
surfers slice mythology.
The pavement swerves
as planned. The lines blur late
no matter which window
the ocean can be seen from.
It's bedtime for all the birds
with a hundred or so miles

to go. The gasman in Bend,
Oregon, asks, How much?
and before I unbuckle, I'm off
to the races, knowing rarely
do we stop in The Dalles.
Having forked in Weed,
we slope past valleys on
the way to our valley, in which
non-citizens make the land
profitable. My great-grands
leased from the Yakama
before the war—so much
war has shaped what I know
and don't know. The country
roads, lined with crates of stone
fruit, never returned the interned,
so a different kind of migrant
ensued. Not knowing the dust
carries on the golden hour, I
trap grasshoppers in mason jars
and walk away because I can.

Dujie Tahat is a Filipino-Jordanian immigrant living in Washington state. They are the author of *Here I Am O My God*, selected for a Chapbook Fellowship from the Poetry Society of America, and *Salat*, selected as a winner of the Sunken Garden Chapbook poetry prize from Tupelo Press and long-listed for the 2020 PEN/Voelcker Award for Poetry Collection. Along with Luther Hughes and Gabrielle Bates, they cohost *The Poet Salon* podcast.

Miriam Alarcón Avila

Through the Lens of My Camera

In 1985, when I was fourteen years old, I survived a devastating earthquake that killed thousands of people in México City. Afterwards, as I wandered through the destroyed neighborhoods, I wished I had a camera to photograph and document what I saw—the disaster itself but especially the human resilience and the generosity of my compatriots willing to sacrifice their lives to help unearth the living and the dead; the survivors who lost everything in three minutes but still found the strength to get up to rebuild their lives again. A camera was a luxury that my widowed mother couldn't afford to buy me. So, I started storing thousands of images in the vault of my memory of the pictures I never took.

From a very young age, I was attracted by shapes and textures, vivid colors, light and shadow. I escaped my reality by gazing at the glare of the sun bouncing off the brightly painted walls and multicolored mosaics of the colonial houses in the historic area of the capital and the murals and stained-glass windows of the Palacio de Bellas Artes.

At sixteen years old, I ran away from home with a journalist ten years older than me, who promised me a camera. It felt like I was prostituting myself for it, but at the pawnshop, when the salesman put the Fujifilm camera in my hands, tears of joy ran down my cheeks. It wasn't the best camera, it was used and beaten up, but it was mine. With it I took hundreds of black-and-white photographs, but I couldn't see what they looked like because I did not have money to pay for the equipment to develop the film. So, I stored them away for years.

I knew that art was my destiny; however, I let myself be misguided by the false story that art is only for the rich, not for those of us who have to earn enough to support ourselves. When selecting my career, I betrayed my soul and chose the safe path. I studied science, and I married a scientist with whom I had two intelligent kids. And since in the world of science, success only comes after a PhD, my husband decided to move us to the United States.

My family and I emigrated from México to Iowa with the dream of obtaining a good education. My goal was to work as a visual artist and find a way to fulfill my dream of becoming a photographer. We arrived in Iowa City at dawn on Martin Luther King Jr. Day in 2002. The January cold penetrated our bones to the marrow. The transition to life in Iowa was shocking. We experienced a whirlwind of emotions, doubts, and confusions, especially with the language, culture, and food, but also with the sudden changes in weather, both in nature and in my new society.

Around the town, while carrying my one-year-old son in my arms and with my little girl holding tight to my skirt, people would look at us weirdly. At hearing me speak, they would look at me up and down and ask, "Are you immigrants?"

I responded with an outright, "No! We are only here to study at the university."

In my mind, there was no possibility that the United States would be the place where I would raise my children. Perhaps, unconsciously, my immediate response was an act of self-preservation. I didn't want anyone to see us as immigrants because with that came the discrimination and anti-immigrant sentiments infecting this country. It was safer to be a foreign student.

A year later, I had the good fortune to win a scholarship to take a black-and-white photography course at the university. My professor, who was Puerto Rican, helped me by letting me attend his photojournalism and photo documentary classes in exchange for helping him in his photography lab. That was when I finally developed all the hundreds of black-and-white photographs of my Mexican past. With many of the photos, I remember perfectly the moment when I pressed the camera shutter; with others, it was like rediscovering pieces of my life, memories that I had completely forgotten.

When my husband reached his doctoral goal, it came at a great cost. In the process of getting his PhD, he had neglected his family, the rigorous program having left him very little time to spend with us, so when the time came to leave Iowa and return to the uncertainty of our native country, my children and I could not bear it. I didn't want to leave my photography dreams unfulfilled. My little seven-year-old begged me, "Mom, please I want to have a good education, please let's stay here, I want to finish school!" When I saw my little girl with tears in her eyes, I made a hard decision.

My husband left us and returned to México to continue his career as a scientist. But things didn't work out the way I'd hoped. Not only were my children and I now undocumented, having lost the visa he'd obtained for our family as a foreign student, but as a single mother with financial challenges, I had no choice but to postpone my studies and pursuit of photography in order to raise my two children. For the next ten years, I did the impossible to give them a good education, holding down a very stressful full-time job to pay the bills. There were times when I felt hopeless because I wasn't achieving my goals as a visual artist, and I was living in a country that didn't want me.

It was hard for my children too. One day, my eight-year-old daughter came home from school crying because her best friend had told her she preferred to play with girls that had white skin, blond hair, and blue eyes. While I comforted my little girl, trying to find the words to apologize for the behavior of "her friend," I could not forgive that girl for being so cruel to my daughter. Still to this day, after several years, I am embarrassed at how much anger I feel when I see that little girl, now a young woman.

In Iowa, I stopped calling myself a Mexican; I became a Latina. I found a new family in Spanish-speaking friends from numerous countries, sharing the cultural similarities of expatriate Latin American populations seeking to find a place in a country that did not recognize our loyalty and commitment. We share the same feeling of invisibility; we are here, but we do not count.

In April 2015, a last-minute decision to escape from my

ordinary life changed things for me—I attended a photography symposium. When I arrived, all my memories from my childhood flashed before me as I heard the inspiring voices of the photographers speaking onstage. I remembered so vividly my desire as a young girl to document, to play with light and texture, to capture images with the camera that I didn't have. I began to cry when I heard and saw the incredible work of these photographers before me. I couldn't help but compare it with my thankless full-time job. A little voice in my head asked me: "Miriam, what happened to you? Why don't you work as hard as these photographers? What happened to your dream of making photographs?"

In the middle of my mental tornado, my kids came to my mind—I couldn't just forget about them. The tears continued rolling on my cheeks, but I tried to hide them. That day I received the best advice anyone had given me. One of the photographers from the symposium asked me what was wrong. I had never shared things about myself with someone I had just met, but that day I did. He probably thought I was completely crazy, but for me, his words became a beacon in the dark. He told me: "Miriam, in four years, your children will be in college and you can work to achieve your goals, so you should make a four-year plan, prepare yourself, and when the time comes, work as hard as you can to achieve your goal."

In September 2015, I started my four-year plan, and in September 2019, my two children were in college. I was free.

Now, I am working very hard to achieve my goal of creating strong visual works of art that could foster a better, sustainable, and inclusive world. Through the camera, I ask myself: Can we educate ourselves, especially our children, to avoid perpetuating negative stereotypes about immigrants?

I started a photo documentary project on the lives of Latino immigrants in Iowa. But even here, I encountered obstacles. The very people whom I sought to showcase were concerned about the intrusive nature of the camera and its ramifications in the US today. "But I don't want them to see me," I was told. "I don't want to be identified."

I knew that I had to find a way to protect the identities of my sources of inspiration, without obscuring or hiding their faces. Our history and our presence in this country have already been hidden for a long time. Instead, the stereotypes of "bad hombres" have dominated the narrative, perpetuating the disparities generated by our brown skin, our language, and our stories.

While driving to work, I remembered my childhood hero "El Santo," a masked Lucha Libre wrestler in México. I remembered my fondness for his films and his spirit of justice, and how I became aware of the significance of the word "lucha," which has a double meaning in Spanish. On the one hand, it is the name of the wrestling match, and on the other, it is the battle we carry out to overcome obstacles—to fight. A "luchador" is one who fights to get ahead, engaged in a struggle to achieve his or her goals. In that moment, remembering the silver mask of "El Santo, el Enmascarado de Plata," I realized that this was the symbol I needed to protect the identities of my interviewees—and, at the same time, empower them by recognizing them as "superheroes."

Instead of hiding their brown faces, I decided to cover

them with glitter, color, and sequins. With each person, I designed and made customized masks that reflected their migratory struggle. I photographed recent immigrants, as well as others who identify with Latino heritage as second- or third-generation residents. I found Luchadores going to Latino festivals or any other event organized for Latinos. We held the interview and took the portraits, either at their home or in a meaningful place for them, and I wrote their history in a poem. I had to learn to separate my emotions and create a good connection between the Luchador, my camera, and me, offering a safe space where they could express their own experiences without fear, for their voices to be heard, bringing them out of the shadows, making them visible. It was powerful to see them transformed into superheroes by wearing their own Lucha Libre mask.

This project has become part of my life's mission as an artist and advocate. While the COVID-19 virus, which has exploded in Latino communities, has slowed my work, "la lucha" continues. I am luchando in my determination to bring these stories into the light, into our communities, and into the "rings" of luchadores across the country. Luchando and fighting to break down the false stereotypes of Latino immigrants in the US.

My camera and I have taken a long journey in this country. But as Che Guevara once said: "La única lucha que se pierde es la que se abandona." The only fight that is lost is the one that is abandoned.

Luchadora con fe todo se puede (2017)

Miriam Alarcón Avila is a visual and multimedia storytelling artist. Born in México City, since the age of sixteen, she has dedicated her artistic pursuit through photography and visual documentation to the goal of creating a legacy that inspires others to shape and support an inclusive and sustainable world. In 2002 she immigrated to Iowa, with the intention of studying photography, where she began working with digital media to document the work of musicians and performance artists, while maintaining the goal of embracing and displaying the cultural diversity of Iowa. In July 2017, Miriam received a grant from the Iowa Arts Council to work on the "Luchadores Immigrants in Iowa" project, a photographic documentary project to give new Iowans a voice and share their challenges as immigrants of Latino descent.

Yosimar Reyes

Silicon Valley, CA

I live with my grandparents in a two-bedroom apartment infested with roaches. We rent out the two bedrooms to indigenous day laborers from Puebla and Oaxaca. My abuelo, Papa Tino, stretches his old body on the living room couch, his feet dangling over the armrest. Abuela and I sleep on the floor covered in thick cobijas San Marcos, each with a design of some exotic animal: a peacock, a leopard, a panther. Abuela runs a soup kitchen, and the laborers crowd our small home telling stories in Mixteco or Nahuatl.

At eleven, I begin teaching basic English sentences to the men while they eat dinner. Most of the sentences are to help them communicate with their bosses and to negotiate pay.

"Ten dollars per hour, se pronuncia," I say.

"Ten dolers per hower," they repeat.

Since I was young, I've known that I have something valuable these adults do not—English. Unlike them, I can maneuver my way around this country and if necessary, defend myself. The combination of being street smart and book smart makes me a force to be reckoned with.

On my twelfth birthday, since there is never any celebration on such occasions, I persuade my abuela to buy me a computer instead. We go down to Fry's Electronics and Abuela has me select the one I want. I know this is huge, and I am probably never going to get another birthday present ever again. Abuela earns a living recycling bottles and cans, so for her to save five hundred dollars for a computer is a big deal. She doesn't like to say that we are poor because she says it is an offense to God. "Con que tengas salud y manos para trabajar siempre dar gracias a dios," she says.

I pick out a Toshiba laptop, and once I get home, I type my first journal entry: "Today is my birthday, and today I start my writing career."

El chisme that I own a computer spreads like fire in my neighborhood.

Back then, an electronics factory located in Fremont, California, operated a huge warehouse where computer products were manufactured. In Silicon Valley, most of these companies subcontracted staffing agencies for dependable workers. Everyone knew in our neighborhood in East San Jose that on the corner of King and Story, past the Don Roberto's Joyeria and Mi Pueblo Supermercado, there was a small office that hired undocumented workers. All you needed to present was your documents. It was also known that this staffing agency had a Don't Ask, Don't Tell policy of sorts. We do not ask you if these papers are fake, and you do not tell us if they are fake.

With no English skills, workers would ask around who could help them write résumés or had access to a computer or printer.

This is when my career as a writer and translator began.

Soon after I get my computer, there is a knock at the door. "Gordo, ¿me puedes ayudar?" Yoli, my next-door neighbor, asks.

She explains that she is applying for a position at the factory, but they are asking her for a résumé. Luckily, my computer came installed with a program that just so happens to have a résumé template.

"A ver, Yoli, dime ¿para qué eres buena?" I ask.

"Ay, no sé, tú ponle lo que quieras," she replies.

I get to work, and by the time I am done, on paper Yoli has become the ideal candidate. I type generic abilities such as *punctual*, *fast learner*, and *responsible*.

"Ay, hasta yo me lo estoy creyendo," Yoli jokes.

It is this small skill, this ability to use words and translate, that positions me in my neighborhood as a resource. Before I know it, I am running my little Educational Opportunity Program (EOP) office for migrants out of my living room.

I write cover letters, translate legal documents, and add my name as a previous employer on their made-up résumés. My neighbors, señores from Puebla, señoras from Oaxaca and Guerrero, knock at my door, all with different needs, knowing that because I speak English, I can help.

My neighbors pay me in soda, in tamales, and in raspados from the street vendor.

"Eres inteligente," they say.

I smile, but deep down inside I know they equate my intelligence to my knowing English and my ability to meet their immediate needs, and their praise is also a way to butter me up for my labor.

On my sixteenth birthday, while most kids at my school are learning how to drive and brag about how their parents are buying them their first car, I can't help but think that God is testing me. Stephanie's mom just bought her a new car for her sweet sixteen. She drives past me down Story Road, honking and waving as I am sweating bullets running to school. You would think she would stop to give me a ride, but no, that would be too much for her; she says, "I would stop, but honestly I got to stop at Starbucks first, and you know the lines are always so long, plus I don't like to be late to class. Besides, you can use the exercise."

Stephanie is the type of citizen I pray to God gets her citizenship revoked. When she tells me, *"Ugh, my parents want to go to México for winter break. You are so lucky your parents can't go anywhere,"* I pray, *"Please, diosito, make someone steal her social."*

As an undocumented kid, my rite of passage is different. My abuela cannot take me to the DMV for my driver's permit. Instead, she makes a phone call.

Abuela dials La Güera, and in coded language over the phone, explains the work that she needs done.

"Necesito un trabajo para mi nieto."

La Güera lives in el apartamento 23. She got her nickname because she is light-skinned, a Mexican con los ojos claros and light brown hair. She mostly keeps to herself, but around the neighborhood, it is an unspoken fact that she's the plug that can get you a mica.

That fake green card meant you could apply for a job, and even though most of our neighbors worked for employers that were well aware they were hiring undocumented workers, it was a formality of sorts to present documents at these staffing agencies, even if they were falsified.

To many, these illicit activities are seen as criminal, but to us, La Güera is a Robin Hood. She is the entryway into the American workforce—no more criminal than employers that salivate at our desperation and exploit our dire necessity to feed our families. Finding a job in this country is hard enough. Most of my tíos wake up at five in the morning and walk to the Home Depot down the street. On my way to school, I wave at them. You see a herd of brown men either playing dice or staying vigilant for possible employers looking for workers. It's sad to see these grown men begging white people for jobs. They rush to the cars, and each in their broken English tries to give a lower price aiming to make enough for next month's rent.

We walk over to el apartamento 23, and La Güera opens the door. She has no furniture in her apartment or has decided to go for a minimalist look. I do not think much of it but later realize she does this in the event that the feds crack down on her operation and she needs to flee.

"Pásele, Abuelita."

My abuela is the oldest one in the neighborhood, which has gotten her some sort of street respect.

"Póngale el nombre y la fecha de nacimiento que quiera."

La Güera instructs us to write down whatever name I want and date of birth on a blank sheet of paper.

Abuela tells me to add two more years to my birth date; that way employers won't hassle me for being an underage worker.

La Güera takes the paper and walks into her room. Forty-five minutes later, she comes out with a laminated ID with my face on it. I look like a kid even though when we went to take the ID pictures at Fotografia Medina, Abuela did everything to try to make me look older. She spiked up my hair and combed the little whiskers sprouting over my lip.

"Ay, no les importa, necesitan trabajadores," Abuela assures herself.

My ID looks fake. I have never seen a mica, but shit, I know this looks like an arts and crafts project. I go home and after the hundreds of résumés I've written, I finally type out my very own. For skills, I make sure to add "bilingual."

Abuela says it's not always going to be like this. I want to believe her, but something tells me it will get a lot worse before it gets any better.

Abuela finds a job for me. She says the electronics factory where Tía Elo works is hiring, and they don't check socials. According to my abuela, this factory gets away with hiring undocumented workers because it is a staffing agency run by Chinos. I tell my abuela the folks are Vietnamese, NOT Chinos, but you know Mexicans and their racism. I mean, come on, Abuela!

It is five in the morning, our shift at the factory starts at six, and since none of us has a driver's license, we have to wait for La Raitera, la señora who is the only one with a license and gives rides to and from the factory in Fremont. She charges fifty dollars a week per person, and being that she manages to squeeze all six of us in her Honda Civic, well, you can see that this homegirl is making hella bank.

We get to work. It's a huge factory and the workers are mostly older señoras and one or two dudes. No one speaks English here even though the line leader commands us in broken English; in my head I correct her English 'cause I'm a bitch like that.

Our job is anything but fun; we stand on our feet for eight hours, sometimes twelve depending on whether we are forced to work overtime. It's an assembly line, so we stand right in front of one another, but we aren't allowed to talk, much less chew gum. We are like human robots assembling computer products unable to have any social interaction.

I stare at las señoras. In my head, I think, *They should build robots to build this shit 'cause this damn work is so mindless.*

No longer the kid helping build cover letters and résumés, I am now one of the workers. Las señoras talk in whispers, crack jokes under their breaths, and it makes the hours go by fast. In the lunchroom, they let out their personalities. Having to repress themselves while on the floor causes them to be wild, funny, and loud in the lunchroom.

I am the youngest, and I stand out. Most of them being mothers themselves, they ask me, "Mijo, ¿por qué estás aquí? Tú tienes que estar en la escuela." Concerned that I might be too young for this job, they ask me about my dreams, about my English, and they tell me, "Este trabajo no es para ti."

I assume they think that because I am young and know how to speak English, this country has endless opportunities for me. I am not sure whether they are aware that I am also undocumented, that like them I presented a crooked ID, that after high school I am not sure who will hire me, that maybe the best I can hope for is that one day I too can become a line leader in this factory.

On my twenty-fifth birthday, a year after DACA was announced, I finally apply. I was hesitant to apply for the program and waited to see what would happen to the first applicants. I imagined the danger of giving the government all my information. At any moment, like the history of Japanese internment in this country, if we registered they could come after us.

In the mail I receive a social security number and a work authorization card. I examine it, and I find myself amazed by all the technology used to produce these documents. They call it biometrics. At the USCIS office, they scanned my fingerprints and took my picture. The office was sterile, and I had to pass through a metal detector and turn off my phone. It took all of that to get me this card, but I know it is real and now no one can deny me opportunities.

Abuela throws her hands up to the sky. "Gracias diosito," she says.

I look at my picture and my number. It's temporary, I know, but I also know the doors will open.

I still do not drive, but I have left the factory job in Fremont behind me and enrolled at San Francisco State University to study English, the very same thing my neighbors praised me for. As I write about the anecdotes I experienced with my neighbors, I can't help but smile now, how silly we looked doing our best to survive in this country. It took all of us to make it. We took the little opportunities given to us and built our lives, poco a poco.

Yosimar Reyes was born in Guerrero, México, and raised in Eastside San Jose. Reyes explores the themes of migration and sexuality in his work. The *Advocate* named Reyes one of "13 LGBT Latinos Changing the World," and *Remezcla* included Reyes on its list of "10 Up and Coming Latinx Poets You Need to Know." His first collection of poetry, *For Colored Boys Who Speak Softly . . .* was self-published after a collaboration with the legendary Carlos Santana. His work has also been published in various online journals and books, including *Mariposas: An Anthology of Queer Modern Latino Poetry* (Floricanto Press), and *Queer in Aztlán: Chicano Male Recollections of Consciousness and Coming Out* (Cognella Press). He is a Lambda Literary Fellow and the recipient of an Undocupoets Fellowship.

Grace Talusan

Counter Encounters

I have always been acutely aware of the power of the person behind the counter. These are the gatekeepers, the enforcers of the policies and rules that the more powerful write. Sometimes they move interminably slow as if to drive home the point that they are in charge and will serve you at their pace, not yours. Sometimes they take long breaks between clients but yell if you don't jump from your seat when your number is called. They comb through the paperwork you have carefully assembled and point out all the ways you've failed—a document is missing or not notarized, you can only pay by check or money order, or no, you cannot pay by check or money order. You'll have to come back and spend another morning waiting your turn.

I am especially sensitive to the power of paperwork and the people who process it because my parents, sister, and I spent years as undocumented immigrants. I had deportation orders at eight years old. My three younger siblings had birthright citizenship, and our immigration lawyer hoped their status as minors would tip things in our favor. It was a strange reversal, the youngest members of my family possessing more power than we did simply by being born here. When President

Reagan signed the Immigration Reform and Control Act of 1986, we began a paperwork process that took years. We became supplicants at the altar of the U.S. Immigration and Naturalization Service.

We waited without complaint, quietly and obediently, for what seemed like days in the molded plastic chairs of government offices for our turn with the often-prickly bureaucrat behind the counter. My father was a different man when he faced that person shuffling through the papers he'd painstakingly gathered. As a well-known physician in our town, with devoted, lifelong patients, my father was accustomed to respect, but facing the counter at the immigration office, he was an ant avoiding a shoe. Everything out of his mouth was "Yes, sir" and "Yes, ma'am," the Filipino ironed from his English until his voice was as smooth as the local weatherman's. If he missed a line on a form or checked the wrong box, he said, "I am so sorry, sir. How careless of me. I hope you can forgive me."

Every step of our naturalization process brought fear and threat, the possibility of denial. Of hearing, *Go back to where you came from; you're not allowed to be here.* And yet, we jumped through every hoop with a warm smile. Isn't that the most important lesson we learn as new Americans? How to grin and bear it? We came to believe in politeness as a shield. We aimed to prove how worthy of respect we were, how good at complying with instructions and following rules. We believed that being friendly and accommodating would protect us from the violence and capriciousness of white supremacy.

Like my father, I was also in terror of the people behind the counter. They often stood between me and something I needed, blocking the next step on my path, and on a whim, could make life easier or harder for me. As a child, I loved reading books but dreaded checking them out from the children's desk at the library. I was afraid of the women dressed in black polyester with their hair coiled tight as fists and a scowl on their faces. I would tremble as I confessed my library crime—a picture book my baby brothers had scribbled over or gnawed on, or more often, an overdue chapter book. The woman behind the counter would say, "Two cents per book per day, not counting days the library is closed, comes to—" while punching the buttons loudly on an oversized calculator. Flattened with shame, I willed myself not to cry.

Sometimes the coins in my palm didn't add up. "You may not check out another book until your fine is paid," she said with righteous pleasure.

"So sorry, thanks," I said as I turned away, red-faced and desperate for air, abandoning the books I had already thought of as mine: *Freckle Juice* by Judy Blume, *Ramona the Brave* by Beverly Cleary, *Amelia Bedelia* by Peggy Parish, and *Betty Crocker's Cook Book for Boys and Girls*, feeling as if I didn't deserve them.

My mother wasn't like my father or me. She didn't cower before the gatekeepers, and she wasn't intimidated by white Americans. In the Philippines, she came from landowners and politicians and brought a sense of entitlement with her to America. In the 1980s, she faced her new country armed with fashion status symbols—blue jeans embroidered with Gloria Vanderbilt's signature and a swan on the pockets, Izod's alligators, and Ralph Lauren's horse rider–and–mallet emblem. She had an ear finely tuned to detect disrespect, and she couldn't resist the opportunity to argue, in the Boston accent that she quickly acquired, with anyone who dared treat her poorly. Like that time at Saks when she asked about the Fendi bags locked behind the glass case and the saleswoman, with keys in hand, scowled and asked, "Are you sure? Do you know how much these are?"

My mother pointed to a Fendi wallet on a tray. Once the saleswoman placed it on the counter, my mother said, "I'll take it," waving her Saks charge card in the air without even checking the price.

Ironically, I became one of the people behind the counter when my friend got me an afterschool job at the library. After I was hired, I showed off, announcing to friends and family that I was a librarian, until my friend corrected me. "You're not," she said. "There's only one librarian there, and he has a master's degree in library science."

Library was a science? Science was biology, chemistry, and physics, wasn't it?

"This is an entry-level job," my friend explained. "You're a clerk. A page."

I was a page, an apt first job for someone who loved books, and my duties included helping patrons find what they were looking for, a book or the bathroom; spending hours shelf-reading the Dewey Decimal call numbers typed onto the white stickers on book spines for anything out of order; gluing plastic sleeves around new books to protect them; and "weeding," the saddest task of all, making a list of books that had not been borrowed for decades. These unread tomes would be plucked from the stacks to make room for newer books that would now have their chance to be loved.

When it was my turn to work in the children's wing behind the glass-topped wooden desk as wide as a raft, the library was still charging fines. The library lent out VHS tapes, with late fees much steeper than those for books. I was always extra friendly when I tallied up a patron's fine. If they didn't have it, I shrugged and said, "That's okay." As the person behind the desk, I had the power to forgive a fine, which made it all the more disgusting to recall how as a child I hadn't been afforded the same grace.

Most of the people I've encountered would not consider themselves racist or prejudiced. And yet, their beliefs about immigrants are often expressed through impatience and contempt, a heavy sigh or eye roll. In the days of paper forms in triplicate, when I've filled them out incorrectly while registering my car or mailing something overseas, I've been asked whether I know how to read. If I pause a beat too long to answer the questions of a person across the counter who wears a uniform, a police officer, or border agent, I've been asked whether I speak English. For a few years after 9/11, I was selected for a random search almost every time I took a flight. Standing on the other side of the table, I'd swallow my anger and nod agreeably as TSA agents rummaged through my roller bag. A good American has nothing to hide. I wanted to be both—good and American—so I complied.

At airports, my skin color makes me too obvious, an easy target. But other times, it renders me invisible. Like the time I went to the mall to return an item during the Christmas shopping season. The white person behind the counter locked eyes with the white person standing in line behind me and offered to help her first.

"I am next," I said, equally angry that the white person behind me rushed to accept.

"I didn't see you," the person behind the counter said with a shrug.

"I know you didn't," I said. "And yet, here I am."

In 2016, when a new president sat behind the most powerful desk of all, one of the first things he did was to sign the Muslim travel ban. He ordered an end to DACA, which, had he succeeded, would have left hundreds of thousands of DREAMers unprotected. The Philippines has the second-highest number of DACA recipients of any Asian country, with more than three thousand. His "zero tolerance" policy separated migrant families with young children, some permanently. His administration denied thousands of US-born Latinos their passports, accusing them of citizenship fraud. And right before leaving office, he finalized his "death to asylum" rule, making it almost impossible for many asylum seekers and refugees to be approved to make the US their home. With every flourish of his pen, he put into motion white supremacist policies that were not only racist and anti-immigrant but cruel and inhumane. Even the verbs to describe these policies are violent:

cut, block, deny, separate permanently. So many times, in the past four years, I've found myself repeating the sentence, "The cruelty is the point."

The morning after Trump's election, I started carrying my passport in my bag whenever I left my home. I was afraid my hard-earned US citizenship might not protect me from the man who had the worst customer service skills of anyone I'd encountered.

Decades after my childhood experiences at the library, I took my young nieces and nephews there for story hour, and I found it unrecognizable with recent renovations and updates. While they were checking out books, I hovered over them. I would not subject them to the same humiliation. I wanted to protect these children for as long as possible from the casual dehumanization that many white people wield and relish in everyday encounters. I was an adult now with years of practice speaking up. I introduced myself to the new librarian, a tall woman with white short hair who smiled warmly at me, and I shared my childhood experiences with the women in black, who seemed to have disappeared from the library like those unread books from the stacks. She waved her hand as if she could push away the bad odor of my words from the air, smiling sheepishly. "Oh, we don't do that anymore. We don't even charge fines. We are a new library."

Maybe one day soon we can say the same thing about the US—We are a new country.

Grace Talusan

Grace Talusan's first book, *The Body Papers*, is a Massachusetts Book Awards winner in nonfiction, a *New York Times* Editors' Choice selection, and the winner of the Restless Books Prize for New Immigrant Writing. She has published essays in the COVID-19 anthologies *And We Came Outside and Saw the Stars Again* edited by Ilan Stavans and *Alone Together* edited by Jennifer Haupt. Her short story "The Book of Life and Death" was chosen for the 2020 Boston Book Festival's One City One Story program and was translated into several languages, including Tagalog. She is the recipient of a US Fulbright Fellowship to the Philippines and an Artist Fellowship Award from the Massachusetts Cultural Council. She was born in the Philippines, arriving in the US with her parents at age two and becoming a US citizen in her twenties. She is the Fannie Hurst Writer-in-Residence at Brandeis University.

Dulce Guerra

Sweet Grass

The roaring sound in the yard made me jump off the couch to look out the window. My dad was pushing a box with wheels across the lawn. Back and forth he went, from one end to the other. I'd never seen a lawnmower. Back at my grandma's house in Huatabampo, Sonora, there wasn't any grass to cut. Her yard, like the roads, was mostly dirt. There was always dust flying around. But here, in the small town of Fair Oaks, Indiana, the flatlands were covered in a velvety green carpet. It was the first thing I fell in love with when we arrived.

The house my dad and uncles were given by the dairy farm where they now worked was surrounded by grass, sugar maple trees, and a post and rail fence in only one corner of the front yard. There was an unattached garage to the right of the house filled with stuff, so we couldn't park a car there. The house was light blue, with white shutters, and had two floors, the upstairs being an attic-turned-bedroom that my parents, sister, and I shared. My parents slept on one side and my sister and I on the other.

The backyard ended with a row of tall pine trees and then a small gravel road that separated the property from a cornfield. The backyard was blanketed with grass nearly as tall as eight-year-old me. I'd arrived in a new place I had no connection to. I not only missed my grandma and my people, but I also missed the land. I missed the soft dirt under my bare feet as I walked outside of my grandma's house. I was now in an unknown place, and everything was so different. For a moment, it hurt in a way I could not put words to. But then, my feet touched grass for the first time, and I knew it would be okay.

I wanted the grass to keep growing, like a selva, my own private jungle. I spent my first two weeks in Indiana playing in the grass, whacking it with a "machete" stick, making teepees out of sticks that fell from the sugar maple trees and covering them with the grass I so loved. I created little imaginary towns, and my very first Polly Pocket dolls were its citizens. Unfortunately, my dad had other plans for my playground. The arrangement for having that house was that he had to maintain the yard.

Now here he was, pushing this strange contraption across the lawn, spitting my joy out on the other side. I opened the window and was hit by an unfamiliar smell—and I realized that the color green has a scent. It comforted me in a way I couldn't explain. I breathed it in greedily. It didn't occur to me that my dad would be moving on to the back, and suddenly it was not that great anymore.

I ran out, bursting through the squeaky screen door.

"Leave the back alone! That's my grass! ¡No lo cortes!" I yelled at him, defending my land. He looked up at me, and I could see in his eyes that he was about to break my heart.

"I have to cut it, mi'ja," he said, kneeling by the lawnmower, checking to see how much gasoline was left in this grass-eating monster.

"No, Papi!" The lawnmower roared to life, and frightened, I ran back inside to the couch to hide, but also to mourn.

My dad had left for Indiana four months earlier. He had worked on dairy farms for most of his life, and my uncles helped get him this new job. My parents decided that he would leave first and get things settled before my mother, sister, and I joined him, but to an eight-year-old girl, I felt that he was abandoning me. My mom reassured me we would see him soon, but I didn't believe her. I became sick, and no doctor could figure out what was wrong with me. I'd wake up to my mom putting cold wet towels on my forehead to bring my fevers down. I had no desire for food nor an interest in playing. I imagined what his new home was like. He'd describe it to me over the phone, but what in the world were fireflies? He'd tell me about all the kinds of ice cream flavors he had seen and which he had tried, butter pecan being his top choice. He'd talk about all the rain and tornado warnings and how one had spun right above the house. He would tell me about the dairy farms and his job as a milker. In all of this, he failed to mention the grass.

The screen door opened with a creak, and my dad came in. I couldn't look at him. How could I? I was feeling the same betrayal I'd felt when he had left me in México.

"Ven aquí," he said.

I thought I was in trouble for raising my voice to him, but I followed him as he led me up the squeaky stairs to the attic-bedroom. He helped me out the window that overlooked the backyard.

"Stay here, don't move," he said.

I sat on the roof of the house, my father warning me not to fall as he walked back down the steps.

"Papi?" I said as his steps faded away.

I was confused and scared. Everything looked different from up here. I could see the gravel road and the fence that protected the field, the bright blue sky, my dad below with the mower. He pulled the string, and it started. I wondered if he was punishing me for yelling at him, forcing me to watch him destroy the only place that brought me joy. But instead of going in straight lines, back and forth, as he had done with the front yard, he started to curve the mower around and made odd lines and shapes. Then suddenly, a *D* appeared. Then a *U*. An *L*. He was writing my name! I watched him, and I wasn't angry or afraid anymore. It was amazing! How did he think of this? Instead of a capital *E* he made a little *e*, and so it read *DULCe*. It looked so silly. When he finished, he turned the mower off, and he came back to the roof and sat with me.

"It looks nice, right?" he said. I couldn't help but smile. I held his hand in approval and in gratitude. "Making the capital *E* would have been hard, so I made it all in one move," he said, tracing the *e* with his finger in the air.

We sat there admiring his work. The wind blew, and the smell of freshly cut grass—the smell of love and hope—graced my face. I took it all in.

The sun was setting, turning the sky into an orange dream. The fireflies emerged from their sleep, brightening the trees with their little lights. The cicadas filled the air with their song.

"It's gonna get dark soon," he said as he helped me up.

"But I want to stay," I said as I held on to his calloused hand.

"Ya no se ve nada, garrapata," he answered.

"Just a little bit longer."

"Ay, Dulcinea del Toboso," he replied, sitting back down, "just a little bit longer or else the mosquitoes are going to eat us up."

I smiled and held on to him and we enjoyed the streaks of light before it was completely dark and my name faded into the night.

"The backyard is now yours because it has your name written on it, BUT, I do have to cut it, understand?"

I nodded. I had to be okay with it. After all, I was now living in a place where grass always grows.

Dulce Guerra was born in Obregón, Sonora, México. In July 2000, at the age of eight, she migrated from Huatabampo, Sonora, to Fair Oaks, Indiana. In June 2012, she married and moved to Los Angeles, California. She currently lives in Palmdale, California, with her husband, Cordero Guerra, their two children, Alora and Benicio, her mother-in-law, and three pets. She graduated from California State University–Northridge in the summer of 2020 with a bachelor of arts degree in Spanish language and culture. She is currently continuing her education at CSUN as a graduate student in the Spanish master's program. She hopes to become a Spanish professor in the future.

Carolina Alvarado Molk

On Paper

When I got home from my naturalization ceremony, I looked at my certificate for a long time. They tell you not to laminate it, not to photograph or frame it, but to simply get home and put it away. I understand why people would want to do more. I understand the instinct to preserve that document in some way that makes you feel like it won't wither away—it's just a piece of paper.

I used to wonder what it would feel like to become a US citizen. I thought it would make me feel safe, free, like I could do anything. I imagined the relief that would wash over me as I walked out of that courtroom—that I would breathe more deeply. It didn't happen quite that way. There's a transition period I hadn't counted on.

Logically, I know that things are different, but it's going to be some time before my body registers that change. The old anxiety is still there, in my sweating palms when I'm asked for an ID, in my accelerated heartbeat when I see a police officer. When I return home from travel, I still think, *This is the time I won't be let in*, and imagine what I would do, where I would go.

When I was younger, all a stranger had to do was ask

me where I was from, and I'd relay my whole life story: my family moved back and forth between the Dominican Republic and the US when I was a child. We settled in Brooklyn for good in 1995. I was open about my status as undocumented—I was bold and defiant and unafraid. Living in New York City, studying and working among immigrants, I trusted that, for the most part, others shared or understood my history.

All of that changed in graduate school. Suddenly, I was an ill-fitting piece in a privileged crowd at an elite institution. "Can't you just apply for citizenship?" administrators asked and frowned in pity or confusion when I explained the unique limitations of my undocumented status. This after they'd tried to rescind my acceptance—someone, somewhere in the process, had assumed I was an international student and had let me in. Semesters ended with emails that asked, "Has your status changed, or are you taking a leave of absence?" To which I'd answer, "Can we consider a third option?" and hope that I'd be allowed to return to campus once more.

Nobody knew what to do with a student like me, and most weren't interested in learning.

It was then that my instincts for self-preservation kicked in, and I went quiet. I let my peers think I wasn't up to the teaching requirement, that I was behind in my coursework instead of undocumented and unable to legally teach for the university. I let them think I didn't care to travel, instead of admitting I wasn't able to. I let them believe I was an overly private and unambitious person. I became private and unambitious.

I wonder now whether my classmates and colleagues knew all along. It would have been difficult not to. There was the time at the liquor store when the clerk rejected my consulate ID. "I don't know what this is," she said, and looked down at it and up at me again, as if waiting for a missing piece of information to land. I didn't have a driver's license, which in New York I could get away with. "A car in New York is a nightmare," I told my classmates. "You don't actually need a license to get around." No one ever said, "But you do need an ID though, don't you?"

There was the prison teaching form I never returned, though I bit my lip and sat through until the end of the info session. At the top was the request for my social security number. "I'll have to bring mine back," I said at the end when they collected the completed forms at the door. All I could think was, *I should've known.*

There was, most obviously, the time at the pizza shop, when the waitress stood at one end of our long table and asked that we pass our IDs down. I watched nervously as my red passport traveled from hand to hand down the table and across to the waitress who made a show of finding my birth date.

I held my breath until she closed it and sat with a hand outstretched as it made its way back to me. Then one of my classmates decided he wanted a look. "Is this yours?" he asked.

I put my hands back down in my lap and smiled what I thought was a casual smile. The table had fallen silent.

"Let's see," he said, and flipped through its conspicuously clean pages, back and forth, back and forth, like a crisp deck of playing cards.

The others looked away, embarrassed and uncomfortable, though it's hard to know if his behavior was the source, or my crimson passport.

"Huh," he said after a long pause, and handed it back to me.

Dinner went on as if nothing had happened—I can't for the life of me remember what it was that we were celebrating. I'm sure I ate and drank and laughed like everyone else. I'm sure I sat there hollow and inwardly trembling.

Often, we regard citizenship as if it were a clean slate, a rebirth. DACA made it possible for me to fulfill my degree requirements, and then a green card obtained through marriage brought me one step closer to safety. I thought citizenship would make me feel whole.

When I think about writing my experience of immigration, I imagine it might feel like reaching back a hand to hold my own. Instead, I often feel stilted. I worry about who I'm writing for. I worry about performing hardship for others. I worry about revealing too much, too little, about getting things wrong.

There is no easy vocabulary for us, for this. There is no vocabulary that can explain what it does to a person to be named "illegal"—before the hashtags, before the "undocumented and unafraid" T-shirts, before newspapers were chastised for calling you "illegal" or an "alien," before it was part of the liberal cause to protest detention centers and to call for the abolishment of ICE. I can tell you that it was terrifying. I can tell you that it was lonely. I can tell you that, as I got older, it got harder and harder to trust people and to connect.

I tell myself that I'm safe now. I can vote. I can travel. I can participate. I can be a part of the country I grew up in in a way I couldn't before. And yet—a piece of paper isn't enough to undo the damage done. Or to soothe the fears that have, for decades, made of my body their home.

I became a citizen in May 2019—two years after I finished my PhD and one year after the arrival of our first child. I remember the sun shining through the windshield as my husband drove us home from the courthouse after the naturalization ceremony. While our son napped in his car seat, we sat quietly—each wrapped in our own thoughts. Erykah Badu played on repeat—the only sound our son would fall asleep to. I looked out of the passenger window and waited for the difference to set in, for the weight to lift.

I've gradually grown more comfortable with my citizenship, and with what it means. I voted for the first time a few months ago. Our small family walked to the drop box near our home, and my husband snapped a picture as I dropped my ballot. We shared the smiling photo on social media, and I celebrated, for the first time, that voting was now something I could do. I'm sure to passersby I looked like any other woman, posing for a picture, voting one more time—but inwardly, something glimmered, something gleamed.

Beginnings, like endings, are bittersweet.

Carolina Alvarado Molk was born in the Dominican Republic and raised in Brooklyn, New York. She holds a dual bachelor's in English and religion from Brooklyn College (CUNY) and a PhD in English from Princeton University. Currently, she is at work on a collection of short stories. She lives and writes in Denver, Colorado.

REYNA'S ACKNOWLEDGMENTS

First, I would like to thank everyone at HarperVia and HarperCollins Español who helped bring *Somewhere We Are Human/Donde somos humanos* out into the world—with special thanks to Tara Parsons, Rosie Black, Juan Milà, Alexa Frank, Ariana Rosado-Fernández, Edward Benítez, Maya Lewis, Sarah Schoof—but above all, to Judith Curr for championing this book from the start. I am eternally grateful to my agent, Johanna Castillo, for her support in every project we've worked on together, but especially with this one because it is so close and dear to our hearts. Much gratitude to my co-editor, Sonia Guiñansaca for working so hard to create safe spaces, including this book, where others like us could speak their truths and share their stories. Thanks for advocating for and empowering our community. As always, my husband, Cory Rayala, copyedits all my projects and this book was no exception. Thank you for your careful edits and suggestions. My deepest thanks to Yaccaira de la Torre Salvatierra for all those late nights in my office and long hours at the coffee shop. Thanks for making the writing hours less lonely. A shout-out to my sister Magloria for her help and support.

With so much gratitude, respect, and appreciation for our contributors. Thanks for joining us on this journey and trusting us with your stories. Your voices have made this book what it is—a powerful testament of the resilience and beauty of our immigrant community. And finally, to Viet Thanh Nguyen for paving the way and being a role model and inspiration to us. What an honor it was to have you as the literary godfather of this anthology!

SONIA'S ACKNOWLEDGMENTS

I arrive at this moment not by my own doing, but through the collective efforts, guidance, nourishment, and love from my community. Yupaychani to all my elders, to fellow Queer/Trans/Non-binary artists, Kichwa diaspora community, and undocumented community. Thank you to my family: Titi, Nube, Rocio, Jesus, Mama Michi, Señora Maria, Tío Angel, Rita, and all my cousins! Special thanks to my siblings: Andy and Erika. Thank you to my parents: Rosa and Segundo Guiñansaca, and my grandparents: Maria Romero, Gerardo Pañora, Cosme Guiñansaca, and Alegria Guiñansaca. My best friend and partner Lylliam, I appreciate all the rough drafts you read through and the pep talks you offered during this process. To my beloved chosen family and friends: Pancho, Little, Bourbon, Turtle, Bambie, Rommy, Leslie, Giselle, Jess, Bianca, Breena, Vivian, Kemi, Emilia, Dorothy, Noemi, my neighbor Sarai, Alan, Monica, Dillon, Phoung, Marléne, Marisol, Daniela, Martin, Jackie, Jose Luis, Jennifer—I appreciate you tremendously. My deepest gratitude to my mentor Ken Chen, and all the brilliant writers that have nurtured my writing: Ruth Forman and Staceyann Chin. To Mr. Murphy, Ms. Greene, Professor Millagros Denis-Rosario, the AFPRL Studies Department, and Women & Gender Studies at Hunter College—you were a blessing when I was an undocumented student looking for a safe space. To the thoughtful and rooted cultural workers and visionaries like Kemi Ilesanmi, Deana Haggag, Alejandra Duque Cifuentes, and Lylliam Posadas—thank you for building out the creative ecosystems we need. My agent, Johanna Castillo, what a journey it has been since that coffee date we had in New York City to talk about the future of my writing and how you wanted to support it. It is special to know I have a fellow Ecuadorian rooting for me. Reyna Grande, my co-editor, we made this happen during a pandemic and in between the hardships of our lives! Thank you for being the Virgo to my Gemini, and for your generous time and impeccable edits. Thank you to Viet Thanh Nguyen for the foreword, the seal the book needed. Thank you to Judith Curr, Tara Parsons, Juan Milà, Rosie Black, Alexa Frank, Ariana Rosado-Fernández, Maya Lewis, Edward Benítez, Sarah Schoof, and the entire HarperCollins editorial team, translators, designers, and all the departments that made this book possible. These books and stories are in urgent need of the support you have offered. What a beautiful step forward, and it is only the beginning. Lastly, the deepest thanks to the contributors who spent the last couple of years working with us, who responded to all of our emails, welcomed feedback, and who trusted us with their pieces. We did it!

ABOUT VIET THANH NGUYEN

Viet Thanh Nguyen's novel *The Sympathizer* is a *New York Times* bestseller and won the Pulitzer Prize for Fiction. Other honors include the Dayton Literary Peace Prize, the Edgar Award for Best First Novel by an American Author from the Mystery Writers of America, the Andrew Carnegie Medal for Excellence in Fiction presented by the American Library Association, the First Novel Prize from the Center for Fiction, a Gold Medal in First Fiction from the California Book Awards, and the Asian/Pacific American Award for Literature from the Asian/Pacific American Librarians Association. His newest book, *The Committed*, the sequel to *The Sympathizer*, was published in March 2021. His other books are *Nothing Ever Dies: Vietnam and the Memory of War* and *Race and Resistance: Literature and Politics in Asian America*, and the bestselling short story collection, *The Refugees. He is the co-author of Chicken of the Sea,* a children's book written in collaboration with his six-year-old son, Ellison. He is also the editor of the anthology *The Displaced: Refugee Writers on Refugee Lives*. Nguyen is the Aerol Arnold Chair of English and Professor of English, American Studies and Ethnicity, and Comparative Literature at the University of Southern California. Most recently he has been the recipient of fellowships from the Guggenheim and MacArthur Foundations and won the Prix du meilleur livre étranger (Best Foreign Book Prize in France) for *The Sympathizer.*

ABOUT THE EDITORS

Reyna Grande is the author of the bestselling memoir *The Distance Between Us* (Atria, 2012; also available as a young readers edition from Aladdin, 2016), in which she writes about her life before and after she arrived in the United States from México as an undocumented child immigrant. The sequel, *A Dream Called Home* (Atria), was released in 2018. Her other works include the novels *Across a Hundred Mountains* (Atria, 2006) and *Dancing with Butterflies* (Washington Square Press, 2009). Her books have been adopted as the common read selection by schools, colleges, and cities across the country. Reyna has received several awards, including an American Book Award, the El Premio Aztlán Literary Award 2006, and the 2015 Luis Leal Award for Distinction in Chicano/Latino Literature. In 2012, she was a finalist for the prestigious National Book Critics Circle Award. Writing about immigration, family separation, language trauma, the price of the American Dream, and her writing journey, Reyna has had her work appear in the *New York Times*, *The Dallas Morning News*, CNN, *The Lily* (published by *The Washington Post*), and Buzzfeed, among others. In March 2020, she was a guest on Oprah's Book Club. Her latest book is the historical novel *A Ballad of Love and Glory* (Atria, March 2022).

About the Editors

Sonia Guiñansaca is an international multidisciplinary artist, cultural strategist, and activist. They write narrative poems and essays on migration, queerness, climate change, and nostalgia, often collaborating with filmmakers and visual artists. They are Kichwa-Kañari, and at the age of five they migrated from Ecuador to the US to reunite with their parents in New York City. In 2007, Guiñansaca came out publicly as undocumented and emerged as a national leader in the migrant, artistic, and political communities where they coordinated and participated in groundbreaking civil disobedience actions. Guiñansaca helped build some of the largest undocumented organizations in the US, including cofounding some of the first artistic projects by and for undocumented writers and artists (Dreaming in Ink writing workshops; UndocuMic, a performance space; and the 2013 UndocuWriting Retreat). They have been awarded residencies and fellowships from Voices of Our Nation (VONA), the Poetry Foundation, the British Council, Creative Time, and the Hemispheric Institute of Performance and Politics. Guiñansaca has been featured by PEN America, *Interview* magazine, *Ms.* magazine, *DIVA* magazine UK, NBC, and PBS and was named one of 13 Coolest Queers on the Internet by *Teen Vogue*. No longer undocumented, Guiñansaca has traveled to London and México City for their migration and cultural equity work, advising on migrant policy and arts programming. They self-published their debut chapbook *Nostalgia and Borders* in 2016. They are a contributor to the new edition of the anthology *Colonize This!* (Seal Press, 2019) and *This Is Not a Gun* (Sming Sming Books/Candor Arts, 2020). Guiñansaca is launching House of Alegria, a publishing house for queer, trans, nonbinary, and migrant undocumented writers.

Here ends Reyna Grande and Sonia Guiñansaca's
Somewhere We Are Human.

The first edition of this book was printed and
bound at LSC Communications in
Harrisonburg, Virginia, April 2022.

A NOTE ON THE TYPE

This novel was set in Bembo, a typeface originally cut by Francesco Griffo in 1495, and revived by Stanley Morrison in 1929. Named for the Venetian poet Pietro Bembo (1470–1547), the font's punch-cut design was a departure from the popular calligraphic style of the day. Its warm touch would inspire later roman typefaces such as Garamond and Times Roman. Morrison, one of the twentieth century's most influential typographers, reworked the typeface for typesetting. Bembo is noted for its legibility and attractive design, making it a popular choice for printed matter.

HARPERVIA

An imprint dedicated to publishing international voices,
offering readers a chance to encounter other lives and other
points of view via the language of the imagination.